SpringerBriefs in Environmental Science

SpringerBriefs in Environmental Science present concise summaries of cutting-edge research and practical applications across a wide spectrum of environmental fields, with fast turnaround time to publication. Featuring compact volumes of 50 to 125 pages, the series covers a range of content from professional to academic. Monographs of new material are considered for the SpringerBriefs in Environmental Science series.

Typical topics might include: a timely report of state-of-the-art analytical techniques, a bridge between new research results, as published in journal articles and a contextual literature review, a snapshot of a hot or emerging topic, an in-depth case study or technical example, a presentation of core concepts that students must understand in order to make independent contributions, best practices or protocols to be followed, a series of short case studies/debates highlighting a specific angle.

SpringerBriefs in Environmental Science allow authors to present their ideas and readers to absorb them with minimal time investment. Both solicited and unsolicited manuscripts are considered for publication.

More information about this series at http://www.springer.com/series/8868

Jaleh Samadi · Emmanuel Garbolino

Future of CO_2 Capture, Transport and Storage Projects

Analysis using a Systemic Risk Management Approach

 Springer

Jaleh Samadi
MINES ParisTech
Paris
France

Emmanuel Garbolino
CRC
MINES ParisTech
Sophia-Antipolis Cedex
France

ISSN 2191-5547   ISSN 2191-5555 (electronic)
SpringerBriefs in Environmental Science
ISBN 978-3-319-74849-8   ISBN 978-3-319-74850-4   (eBook)
https://doi.org/10.1007/978-3-319-74850-4

Library of Congress Control Number: 2018930142

Printed on acid-free paper

This Springer imprint is published by Springer Nature
The registered company is Springer International Publishing AG
The registered company address is: Gewerbestrasse 11, 6330 Cham, Switzerland

Preface and Acknowledgements

The current book is an update of a Ph.D. thesis made in MINES ParisTech, from 2009 to 2012. The research question came up at that time is still topical. That is why we decided to readdress the question and analyze the evolution of the situation concerning Capture, Transport and Storage of CO_2 projects.

I wish to express my gratefulness to all the persons who made this possible, and especially my parents for their endless love and support.

Paris, France   Jaleh Samadi

Contents

About the Authors

Jaleh Samadi was awarded a Ph.D. in Engineering Science at MINES ParisTech, France. Since then, she has worked as Project Manager and Safety Engineer at EReIE (Energy Research, Innovation & Engineering) in France. She has been specially involved in the development and construction of an innovative biogas treatment/bioLNG production unit.

Since 2016, Dr. Samadi continues her professional career as a Project Manager and Business Developer in JIFMAR Offshore Services in France.

She continues her research on the development of CTSC technology as well as Risk Management approaches.

Emmanuel Garbolino was awarded a Ph.D. in Geography at the University of Nice-Sophia Antipolis, and since 2002 has been a Lecturer and Assistant Professor at the CRC, MINES ParisTech.

Dr. Garbolino's research areas include climate change impacts on ecosystems and human societies, modeling of natural and anthropogenic systems, risk engineering dedicated to natural and anthropogenic hazards (risk assessment and prevention, crisis management, and damage assessment).

Dr. Garbolino is a member of the Education and Research Centre on CO_2 Capture, Transport and Storage.

Abbreviations

atm.	Atmosphere (pressure unit of measurement)
Ar	Argon
AS/NZS 4360: 2004	Australian/New Zealand risk management standard, version 2004
Bar	Pressure unit of measurement
BLEVE	Boiling Liquid Expanding Vapor Explosion
°C	Degrees of Celsius (temperature unit of measurement)
CCS	CO_2 Capture and Storage
CH_4	Methane
CO	Carbon monoxide
CO_2	Carbon dioxide
CTSC	Capture, Transport and Storage of CO_2
DNV	Det Norske Veritas
EIA	Environmental Impact Assessment
EOR	Enhanced Oil Recovery
ESD	Emergency Shut Down
EU	European Union
GCCSI	Global CO_2 Capture and Storage Institute
Gt	Giga (10^{12}) tonnes
H_2	Hydrogen
H_2S	Hydrogen Sulfide
HSE	Health, Safety and Environment
ICPE	Installation Classée pour la Protection de l'Environnement
IEA	International Energy Agency
IEC 60300-3-9: 1995	International Electrotechnical Commission standard for risk management. Guide to risk analysis of technological systems, version 1995
IPCC	Intergovernmental Panel on Climate Change
IRGC	International Risk Governance Council

ISO/IEC 73: 2002	International standard for risk management—Vocabulary —Guidelines for use in standards, version 2002
km^2	Square kilometer
km	Kilometer
LNG	Liquified Natural Gas
LSIP	Large-Scale Integrated Project
m	Meter
max.	Maximum
MIT	Massachusetts Institute of Technology
Mtpa	Million tonnes per annum
N_2	Nitrogen
NGO	Non-governmental Organization
NO	Nitrogen monoxide
NO_2	Nitrogen dioxide
O_2	Oxygen
ppm	Parts per million
SO_2	Sulfur dioxide
STAMP	Systems-Theoretic Accident Model and Processes
STEL	Short-Term Exposure Limit
STPA	Systems-Theoretic Process Analysis
t	Tonnes
UK	United Kingdom
USA	United States of America

Introduction

Dear Reader,

Many thanks for choosing this book.

If you think that reading scientific works is usually boring, we are going to make it interesting together.

If you are ready, let's start our journey with some questions. The idea is to provide you with the key notions of this book and give you the desire to continue reading.

Don't worry, the answers are also provided to give you an initial idea about the subject of the book. If you are interested to know more about different topics, we invite you to have a look at the references at the end of each chapter.

Ready for the first question?

Let's go!

Question 1: Do you know what is **Climate Change**?

Answer 1: **Climate Change** is recognized as *an urgent and potentially irreversible threat to human societies and the planet* in the last conference on climate change held in Paris at the end of 2015 (COP21 2016).

Scientists believe that the temperature of the earth's surface is increasing, mainly because of anthropogenic greenhouse emissions, which have been growing exponentially since the beginning of the Industrial Age.

Question 2: What about the **certainty** of Climate Change?

Answer 2: Debates are still ongoing. But a number of scientific studies over the last decade **confirm the certainty** of happening climate change (Wennersten et al. 2015).

Question 3: What are the **commitments** of the latest international agreement on tackling Climate Change?

Answer 3: To meet the objectives of latest international agreement (COP21 2016), all countries should participate to reduce the global greenhouse gas emissions. The global temperature increase should be kept below 2 °C and efforts should be made to limit it to 1.5 °C.

Question 4: Is there any **method** to deal with Climate Change?

Answer 4: IPCC (Intergovernmental Panel on Climate Change) classifies the technical solutions to tackle Climate Change in two categories: mitigation and adaptation solutions (IPCC 2014). However, other classifications are also available. For example, Climate Control methods (such as CO_2 Capture, Transport and Storage or geoengineering), and alternative methods of energy production (such as nuclear and renewables) (Poumadère et al. 2011). These methods could be compared to the IPCC mitigation and adaptation solutions respectively.

Question 5: How we can choose the **most efficient method** to take action on Climate Change issues?

Answer 5: IPCC states that *neither mitigation nor adaptation alone can avoid climate change impacts* (Wennersten et al. 2015). So, there is not a single answer to this question. All solutions are welcome to make the objectives happen. A combination of solutions seems to be the most efficient.

Question 6: Will we talk about all different solutions in this book?

Answer 6: In this work, we put the emphasis on Capture, Transport and Storage of CO_2, as a mitigation solution to deal with Climate Change.

Question 7: What does "**Mitigation**" means?

Answer 7: *"**Mitigation**", in the context of climate change, is a human intervention to reduce the sources or enhance the sinks of greenhouse gases* (IPCC 2014). Capture, Transport and Storage of CO_2 (CTSC) is so considered as a mitigation technology.

Question 8: What is Capture, Transport and Storage of CO_2 (**CTSC**)?

Answer 8: **CTSC** consists of a chain of processes to collect or capture a CO_2 gas stream, transport the CO_2 to a storage location, and inject it into that location.

Question 9: How is **CO_2 emitted** into the atmosphere?

Answer 9: The most significant source of CO_2 emissions is the combustion of fossil fuels such as coal, oil and gas in power plants, automobiles, and industrial facilities. Chemical, metallurgical, and mineral transformation processes, agricultural activities, transportation, burning fuels for heat in buildings, or cooking in homes are some other sources of global greenhouse gas emissions (EPA 2016).

Question 10: What about the **concentration of CO_2** in the atmosphere?

Answer 10: CO_2 is the main greenhouse gas responsible for global warming. The current concentration of CO_2 in the atmosphere is now around 400 ppm (parts per million). Atmospheric CO_2 could reach 500 ppm by 2050 and 800 ppm by 2100 if current rates of greenhouse gas emissions continue (Wennersten et al. 2015). Figure 1 shows the evolution of CO_2 average concentration.

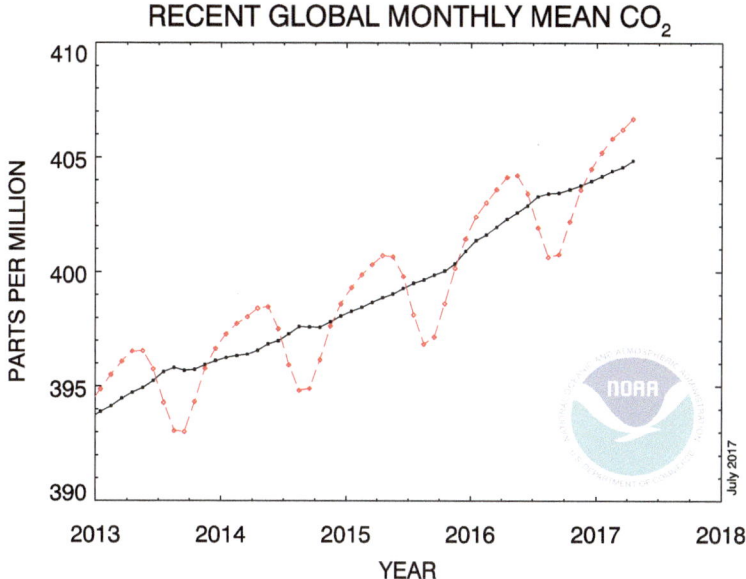

Fig. 1 Atmospheric CO_2 emissions evolution (ESRL 2017) Red line: Monthly mean values Black line: Monthly mean values, after correction for the average seasonal cycle

Question 11: Is there a global agreement about the **limit of CO_2 concentration** in the atmosphere?

Answer 11: There is not a general agreement about the "safe" limit of CO_2 concentration in the atmosphere. Staying under 350 ppm is just a figure which is noted in some scientific publications (Wennersten et al. 2015).

Question 12: **How much CO_2** has been already produced and emitted into the atmosphere?

Answer 12: In 2014, global CO_2 emissions reached 32.4 $GtCO_2$. No surprise, China (28%) and the United States (16%) are at the top of the list of emitting countries. CO_2 emission rate of top ten emitting countries in 2014—which produced two-thirds of global CO_2 emissions—is presented in Fig. 2.

Question 13: Does everyone have the **same idea** about the efficiency of CTSC technology in Climate Change tackling?

Answer 13: CTSC is still an unknown technology for many people even the stakeholders. Some other climate control or mitigation methods like geoengineering are also in the same case (Wennersten et al. 2015; Poumadère et al. 2011).

Perceptions of stakeholders on the effectiveness of CTSC are different. Although most of governments and industries intend to invest in the technology, others such as local communities and NGOs are worried.

Fig. 2 Top ten emitting
countries in 2014 (IEA 2016)

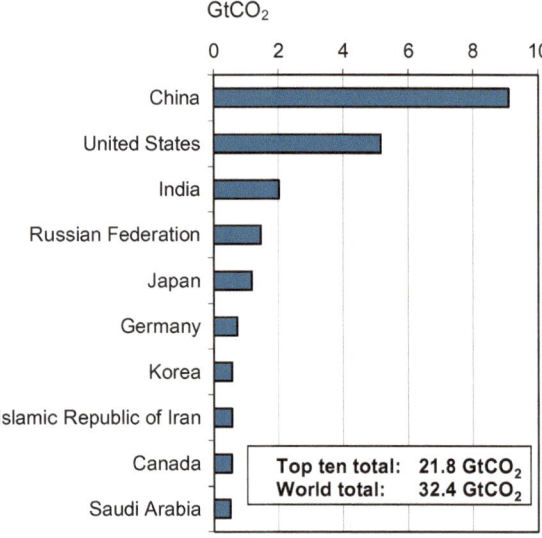

Question 14: What are the **concerns of the stakeholders** about CTSC technology development?

Answer 14: Stakeholders like local communities and NGOs are worried about long-term risks and reliability of CO_2 storage. CO_2 leakage is the most significant concern of these groups since it could lead to risks for human beings, animals, and plants as well as potable water networks.

Question 15: Is there any **solution** to help the stakeholders dealing with their concerns?

Answer 15: What we propose is a **development of adequate Risk Management methods** and use these methods from the very first phases of a CTSC unit development.

We believe that Risk Assessment and Management are essential parts of CTSC development in order to provide answers to the uncertainties and assure the control of well-understood parts of CTSC processes.

Experts' general opinion confirms that Risk Assessment is *vital for the success* of any CTSC project (Wennersten et al. 2015).

An efficient communication process is also required to exchange information about technical, economic feasibility, and social acceptance of the technology.

Question 16: Is there any **risk assessment method available** for CTSC?

Answer 16: Several studies have been carried out on risk assessment of Capture, Transport and Storage technologies. Risks of CO_2 Capture and Transport are supposed to be well understood. Therefore, classical methods have been usually applied for analyzing risks of Capture and Transport subsystems. However, CO_2

storage is known as a *"non-engineered"* part of the process, dealing with various uncertainties (Koornneef et al. 2012). Consequently, most of the available risk assessment studies are focused on CO_2 storage technical aspects of risk.

Question 17: How we can **improve** the available risk assessment methods?

Answer 17: What is neglected in most of the available approaches is that CTSC is a **complex sociotechnical system** for which risks could not be analyzed individually, without taking the whole context into account.

So, our proposition is to take this fact into account and study the whole system of risks associated with CTSC.

Question 18: What is a **Complex System**?

Answer 18: A Complex System is a *system composed of many parts that interact with and adapt to each other. In most cases, the behavior of such systems cannot be adequately understood by only studying their component parts. This is because the behavior of such systems arises through the interactions among those parts* (IRGC 2010).

Question 19: What is a **Sociotechnical System**?

Answer 19: A Sociotechnical System is a one which consists of a technical part which is in interaction with a social part.

Question 20: Is there any **major question about the development of CTSC** projects?

Answer 20: Risks associated with CTSC are not limited to technical risks. Along with technical challenges, CTSC is faced with uncertainties concerning development up to commercial scales. Seventeen large-scale CCS projects are currently in operation around the world (GCCSI 2017). In 2016, forty-three projects were announced canceled or on hold. Financial reasons are frequently noted as the reason for project failure. However, Public Opposition, Legal, Technical, and Policy concerns are some other reasons of projects' cancelation (MIT 2016).

Therefore, a major question about CTSC at the current scale of development is **what are the factors explaining the success or failure of CTSC projects in different contexts?**

Question 21: Any proposition for replying to the here-above question?

Answer 21: In order to answer this question, we propose **a systemic risk management framework** based on the concepts of System Dynamics and STAMP (Systems-Theoretic Accident Model and Processes), developed at Complex Systems Research Laboratory of Massachusetts Institute of Technology.

Question 22: Where does the idea of this book come from?

Answer 22: Aside from the sociotechnical complexity of CTSC system, the idea comes from **systemic and dynamic characteristics of risk**. Systems are regularly adapting themselves to perturbations. Nevertheless, positive feedbacks lead to system destabilization by amplifying the perturbations. So, it is important to identify feedback dynamics involved in the system in order to *better anticipate when risks might emerge or be amplified* (IRGC 2010).

In this book, systemic modeling is proposed as a decision-making support, which provides the grounds of thinking about the components of a potentially successful CTSC project. Each stakeholder is assumed as a "controller", who is responsible for maintaining safety constraints. Safety control structures are developed for several case studies to formalize the relations of stakeholders in maintaining safety constraints.

References

COP21 (2016) Report of the Conference of the Parties on its twenty-first session, held in Paris from 30 November to 13 December 2015, Part two: Action taken by the Conference of the Parties at its twenty-first session, United Nations Framework Convention on Climate Change, 29 January 2016

EPA (2016) United States Environmental Protection Agency, Global Greenhouse Gas Emissions Data. https://www3.epa.gov/climatechange/ghgemissions/global.html#three. Accessed 26 March 2016

ESRL (2017) Earth System Research Laboratory Global Monitoring Division, Trends in Atmospheric Carbon Dioxide. https://www.esrl.noaa.gov/gmd/ccgg/trends/global.html. Accessed 30 July 2017

GCCSI (2017) Large-scale CCS facilities Project Database. https://www.globalccsinstitute.com/projects/large-scale-ccs-projects. Accessed 6 October 2017

IEA (2016) International Energy Agency, CO_2 Emissions from Fuel Combustion Highlights (2016 Edition)

IPCC (2014) Climate Change 2014: Mitigation of Climate Change, Working Group III Contribution to the Fifth Assessment Report of the Intergovernmental Panel on Climate Change, Edenhofer O et al., Cambridge University Press, Cambridge, United Kingdom and New York, NY, USA.

IRGC (2010) The Emergence of Risks: Contributing Factors, Report of International Risk Governance Council, Geneva, 2010, ISBN 978-2-9700672-7-6

Koornneef J, Ramirez A, Turkenburg W, Faaij A (2012) The environmental impact and risk assessment of CO_2 capture, transport and storage—An evaluation of the knowledge base, Progress in Energy and Combustion Science 38 (2012) 62–86, doi: https://doi.org/10.1016/j.pecs.2011.05.002

MIT (2016) Massachusetts Institute of Technology, Cancelled or Inactive Projects. http://sequestration.mit.edu/tools/projects/index_cancelled.html. Accessed 4 July 2017

Poumadère M, Bertoldo R, Samadi J (2011) Public perceptions and governance of controversial technologie to tackle climate change: nuclear power, carbon capture and storage, wind, and geoengineering, John Wiley & Sons, WIREs Climate Change 2011 doi: https://doi.org/10.1002/wcc.134

Wennersten R, Sun Q, Li H (2015) The future potential for Carbon Capture and Storage in climate change mitigation—an overview from perspectives of technology, economy and risk, Journal of Cleaner Production 103 (2015) 724–736

Chapter 1
CTSC, Risk Management and Requirement of a Systemic Approach

Abstract In this chapter, the position of Capture, Transport and Storage of CO_2 (CTSC) in climate change mitigation is introduced. Then, an overview of CTSC projects in the world is presented. Risks associated to each subsystem and the whole chain are presented. Risks related to the whole chain in a complex sociotechnical framework are classified in eight groups: Technical, Risks related to Project, Social, Policy/Strategy, HSE, Regulatory, Organizational/Human and Financial/Economic. Notions of risk and risk management are also introduced, followed by a general recall of the evolution of risk management approaches. In the last part, available risk management approaches for Capture, Transport and Storage are reviewed individually. Finally, integrated (systemic) approaches are argued as essential need of risk management for CTSC.

In this chapter, we will introduce CTSC (Capture, Transport and Storage of CO_2), the risks associated with this innovative technology, and the gaps in available risk management approaches.

This chapter is divided into six major parts. In the first two sections the contribution of CTSC to climate change and its current status in the world are provided.

In the third part, a review of risks associated with CTSC subsystems and the whole chain are presented.

The evolution of risk management approaches is the subject of the fourth part. Limitations of classic methods and the requirement of novel approaches for innovative technologies are discussed in this section.

In the fifth section of this chapter, available risk management methods for CTSC are reviewed.

The necessity of developing an integrated approach is discussed at the final section.

Please note that:

- In this work, "CO_2 storage" refers to the storage in geological formations. Otherwise, the storage system is clearly specified.
- In this work, "CTSC" is used for the integrated chain of Capture, Transport and Storage of CO_2. In a number of citations, "CCS" is referred to the same integrated system.

J. Samadi and E. Garbolino, *Future of CO2 Capture, Transport and Storage Projects*, SpringerBriefs in Environmental Science, https://doi.org/10.1007/978-3-319-74850-4_1

1

1.1 CTSC and Climate Change

CTSC refers to the chain of processes used to collect or capture a CO_2 gas stream, transport the CO_2 to a storage location and inject it into that location. An overall view of CTSC possible systems is illustrated in Fig. 1.1.

The most significant source of CO_2 emissions is the combustion of fossil fuels such as coal, oil and gas in power plants, automobiles and industrial facilities. Chemical, metallurgical, and mineral transformation processes, agricultural activities, transportation, burning fuels for heat in buildings or cooking in homes are some other sources of global greenhouse gas emissions (EPA 2016).

Capture, Transport and Storage of CO_2 (CTSC) is one of the contribution options for mitigating industrial CO_2 emissions in the atmosphere. CTSC technology is developing along with other low carbon technologies such as renewable resources, increasing energy efficiency, fuel switching and nuclear. The set target is halving the emissions by 2050 (compared to the current amount) (GCCSI 2011). The current (April 2017) amount of CO_2 in the atmosphere is equal to 406.67 ppm (ESRL 2017). Key technologies contribution to CO_2 emission reduction is illustrated in Fig. 1.2 (IEA 2017).

There is not a mutual agreement about the necessity and effectiveness of CTSC in global energy policies. Non-Governmental Organizations (NGOs) are major opponents of CTSC development. An example is Greenpeace, which is an international environmental NGO. Greenpeace believes that CTSC is not ready to save

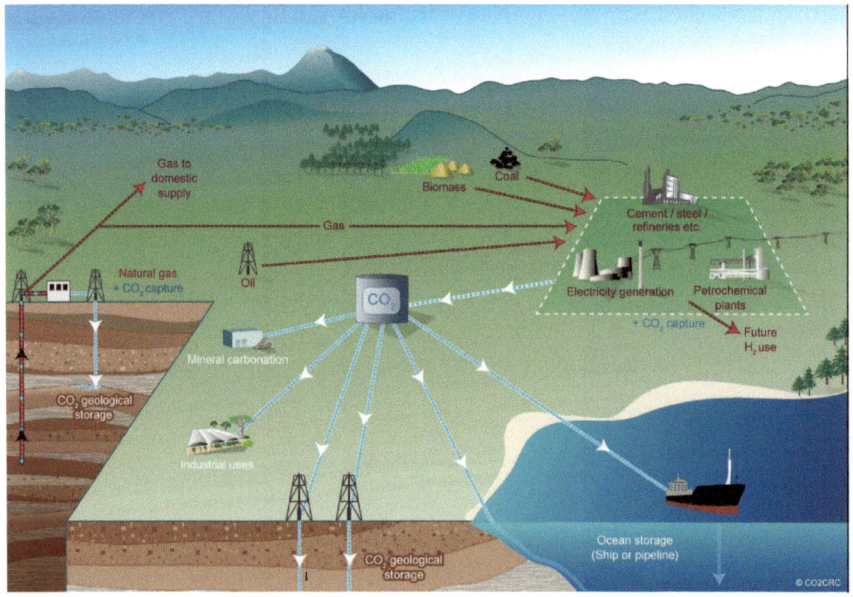

Fig. 1.1 Possible CTSC systems (IPCC 2005)

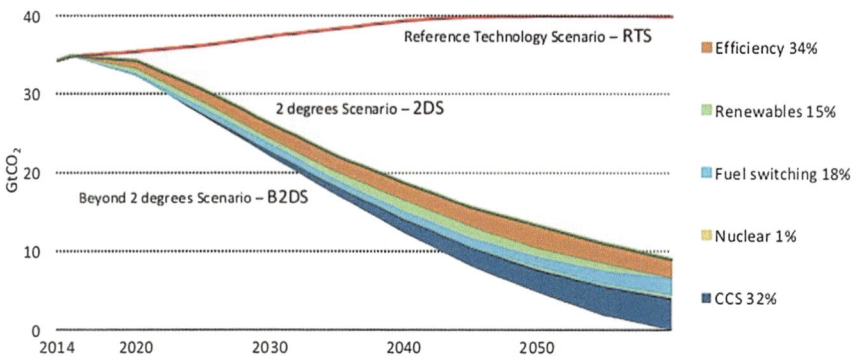

Fig. 1.2 Key technologies contribution to CO_2 emission reduction (IEA 2017)

the climate in time. According to the United Nations Development Program (UNDP), CTSC *will arrive on the battlefield far too late to help the world avoid dangerous climate change* (UNDP 2007). Energy waste, risk of CO_2 leakage, expensiveness and liability risks are some other points noticed by Greenpeace for supporting the idea of conceiving CTSC as "*False Hope*". Greenpeace believes that renewable energy and improving energy efficiency are safe and cost-effective for the climate change problem (Rochon et al. 2008).

1.2   CTSC Projects Current Status in the World

Currently, thirty-seven Large Scale Integrated Projects (LSIP) are identified all around the world (GCCSI 2017a). Global CCS Institute (GCCSI) defines LSIP as the projects which involve all the three subsystems (Capture, Transport and Storage), at a scale of not less than 800,000 tonnes/year of CO_2 for a coal-based power plant and at least 400,000 tonnes/year of CO_2 for other industrial plants (GCCSI 2017b).

In 2012, seventy-five projects have been announced by GCCSI (2012).

These figures show that the number of LSIP projects have been noticeably reduced in the last five years.

The current status of LSIP CTSC projects is summarized in Table 1.1 and could be compared to the status on 2012 (Fig. 1.3).

In 2016, forty-three projects were announced canceled or on hold. Most of the projects are cancelled or put on hold because of economic or financial issues. The reason of cancellation for some projects are not clearly defined. However, Legal, Technical and Policy concerns are some other origins of project cancellation (MIT 2016).

We will not develop the Capture, Transport and Storage technologies in this work. In the next pages, the risks associated to CTSC are discussed.

Table 1.1 LSIP CTSC projects by region and project phase (GCCSI 2017a)

Region	Phase				Total
	Early development	Advanced development	In construction	Operating	
North America	0	2	2	12	16
Europe	2	1	0	2	5
China	6	1	1	0	8
Middle East	0	0	0	2	2
Rest of the world[a]	3	1	1	1	6
Total	11	5	4	17	37

[a]Includes Australia, South Korea and Brazil

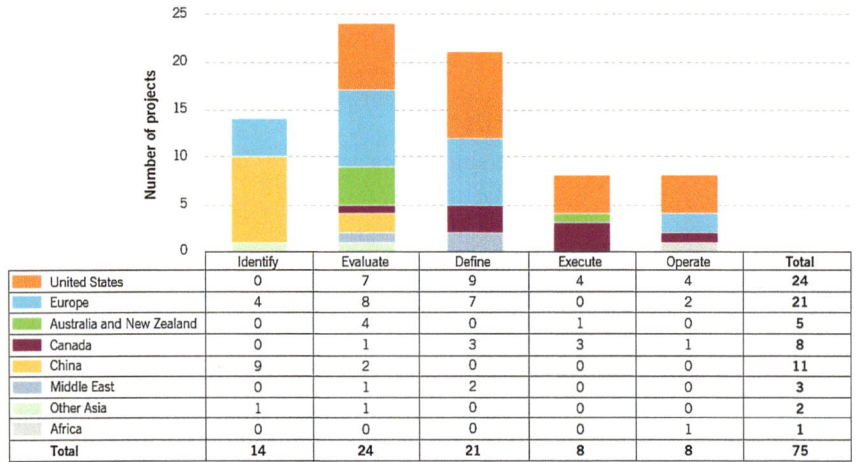

Fig. 1.3 LSIP CTSC projects by region and project phase (GCCSI 2012)

1.3 CTSC Technology and Risks

In order to understand why a systemic risk management framework is required for CTSC chain, CO_2 properties and potential risks are presented here and the risks of CTSC activities are reviewed.

1.3.1 Health and Safety Aspects of Exposure to CO_2

Carbon dioxide is a colorless, odorless, harmless, non-flammable gas (at normal temperature and pressure, i.e. 20 °C and 1 atm.). CO_2 is a constituent of the

Table 1.2 Occupational exposure standards for CO_2 (IPCC 2005)

	Time-weighted average (8 h/day, 40 h/week)	Short-term exposure limit (15 min)	Immediately dangerous to life and health
OSHA permissible exposure limit[a]	5000 ppm (0.5%)		
NIOSH recommended exposure limit[b]	5000 ppm (0.5%)	30,000 ppm (3%)	50,000 (5%)[d]
ACGIH threshold limit value[c]	5000 ppm (0.5%)		

[a]OSHA: US Occupational Safety and Health Administration (1986)
[b]NIOSH: US National Institute of Occupational Safety and Health (1997)
[c]ACGIH: American Conference of Governmental Industrial Hygienists
[d]Corrected based on http://www.cdc.gov/niosh/idlh/124389.html, accessed June 19, 2012

atmosphere and a necessary ingredient in the life cycle of animals, plants and human beings. In addition, there are large amounts of CO_2 in the ocean, about 50 times of atmospheric amount of CO_2 (Johnsen et al. 2009; Serpa et al. 2011).

According to the standards, a concentration of 0.5% is acceptable for a continuous exposure to CO_2, while it will be dangerous if the concentration is more than 5%. Occupational exposure limits for CO_2 are summarized in Table 1.2.

According to DNV, incidents related to CO_2 could be categorized in three main groups: Fire extinguisher systems, Pipelines and Natural outgassing of CO_2 (Johnsen et al. 2009).

The reader is referred to the DNV report (Johnsen et al. 2009) for more information on incidents details. A list of CO_2 vessel ruptures is also available in the same report.

1.3.2   CTSC: Risks Associated to Each Phase and to CTSC Chain

De Coninck et al. believe that the risks of CTSC are difficult to identify, not only technically but due to the stakeholders' different perceptions of risks. Perceptions of energy policy and requirement of low-carbon energy could also affect the perceptions of CO_2 storage risks (De Coninck et al. 2009).

In this part, we firstly summarize the risks related to each phase. Afterwards, the risks of CTSC whole system are discussed.

1.3.2.1   Risks Associated to CO_2 Capture

The most fundamental risks in CO_2 capture processes are associated with the vent gas produced from the capture plant, as well as liquid and solid wastes.

The captured CO_2 stream may contain impurities which would have practical impacts on CO_2 transport and storage systems and also potential health, safety and environmental impacts. SO_2, NO, H_2S, H_2, CO, CH_4, N_2, Ar and O_2 are the impurities that will be available in the CO_2 stream, depending on the capture process type. Moisture of CO_2 from most capture processes has to be removed to avoid corrosion and hydrate formation during transportation (IPCC 2005). Problems of impurities will be readdressed in the next parts.

Another major concern about CO_2 capture is the cost of capture technologies (GCCSI 2011). Several research and development studies are carrying out to find the cost reduction methods.

IPCC believes that *monitoring, risk and legal aspects associated with CO_2 capture systems appear to present no new challenges, as they are all elements of long-standing health, safety and environmental control practice in industry* (IPCC 2005).

CO_2 capture and compression processes are listed as gas processing facilities in several governmental, industrial and finance guidelines. Typical engineering design, commissioning and start-up activities associated with petrochemical facilities are applicable to CO_2 capture and compression. For example, HAZard OPerability (HAZOP) studies are conducted on a routine basis for new facilities (IPCC 2005).

1.3.2.2 Risks Associated to CO_2 Transport

Risks related to CO_2 transportation obviously depend on the transportation mode and on the local topography, meteorological conditions, population density and other local conditions. However, carbon dioxide leaking from pipelines or other modes of transportation could result in potential hazards for human beings and ecosystem. Therefore, public acceptance is a critical issue in large scale development of CO_2 pipelines (IPCC 2005).

Leakage is defined as the main safety issue for CO_2 pipelines in some research studies. *Significant quantities of other components in the CO_2 may affect the potential impacts of a pipeline leak or rupture.* De Visser et al. specified the following Short Term Exposure Limits (STEL) and maximum recommended level of impurities in the CO_2 stream (STEL: Maximum allowed exposure limit for a period of 15 min without adverse health effects). Typical CO_2 volume concentration transported by pipeline is over 95%. For the figures of Table 1.3, the authors

Table 1.3 Maximum and recommended level of impurities in CO_2 from a health and safety point of view (De Visser et al. 2008)

Component	STEL (ppm)	Maximum level (not corrected) (ppm)	Safety factor	Recommended maximum level (ppm)
CO_2	10,000	1,000,000	–	–
H_2S	10	1000	5	200
CO	100	10,000	5	2000

set a concentration of 100% for CO_2 as the reference to define the levels of H_2S and CO (De Visser et al. 2008).

Corrosion is another major problem associated to CO_2 pipelines. To minimize the corrosion, impurities such as hydrogen sulphide or water have to be removed from the CO_2 transported stream. Selecting corrosion-resistant materials for pipelines is also important to avoid corrosion. Corrosion rate, risk of hydrate formation and risk of water freezing will increase in the presence of free water. The amount of free water should be maintained below 50 ppm (Serpa et al. 2011). Other experts propose different limits for water concentration. The limit for De Visser et al. is 500 ppm. Corrosion rates are in the order of mm/year in case of free water presence and in the order of μm/year when CO_2 is dry (De Visser et al. 2008; Seiersten 2001).

Impurities could also change the thermodynamic behavior of the stream. As a result, velocity and pressure drop in the pipeline are subject to change; and transport cost will change accordingly (Serpa et al. 2011). Two phase flow could lead to the damage of compressors and other equipment, and hence should be avoided.

Existing gas pipelines are widely used for CO_2 transportation. The main problems of the existing pipelines are the adequacy of design pressure and remaining service life. CO_2 pipelines normally operate in 85–150 bar, while natural gas pipelines operation pressure is below 85 bar. A great number of existing pipelines have been in service for 20–40 years (Serpa et al. 2011).

1.3.2.3 Risks Associated to CO_2 Storage

There are two types of risks concerning geological storage of CO_2, "local risks" and "global risks". Risks for human beings, animals and plants above ground, contamination of potable water, interference with deep subsurface ecosystems, ground heave, induced seismicity, and damage to mineral or hydrocarbon resources are some examples of local risks (BRGM 2005).

IPCC has categorized the local risks almost the same as BRGM in three groups (IPCC 2005):

- *Direct effects of elevated gas-phase CO_2 concentrations in the shallow subsurface and near-surface environment*
- *Effects of dissolved CO_2 on groundwater chemistry*
- *Effects that arise from the displacement of fluids by the injected CO_2*

GCCSI argues that CO_2 storage will not have an impact on surface water resources, since the groundwater production occurs in depths of zero to 300 m, while CO_2 will be stored at more than 800 m (GCCSI 2011).

"Global risks" refer to the release of CO_2 in the atmosphere, which brings the initial objective of CO_2 storage (reducing atmospheric CO_2 emissions) into question.

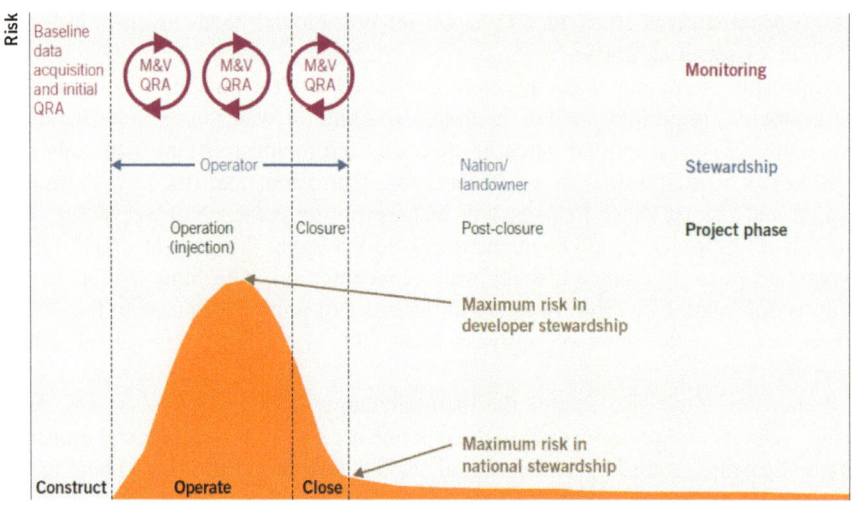

Fig. 1.4 Schematic risk profile for a storage project (GCCSI 2011). Source: Wright (2011), based on InSalah project. M&V: Monitoring and Verification, QRA: Quantitative Risk Assessment

Impurities such as H_2S, SO_2 and NO_2 could increase the risks. For instance blow-outs containing H_2S are more toxic than blow-outs containing CO_2. The acid generated from the dissolution of SO_2 in groundwater is stronger than carbonic acid formed by dissolution of CO_2 (IPCC 2005).

Figure 1.4 illustrates that risks during the lifecycle of a CO_2 storage project are at the highest level near the later stages of injection (Wright 2011). The profile is similar to the one presented by Benson (2007). Risk reduction over time occurs due to the *pressure dissipation and residual trapping of CO_2 in the pore spaces* (GCCSI 2011).

1.3.2.4   Risks Associated to CTSC Whole Chain

In addition to risks related to each subsystem of CTSC chain, it is essential to analyze the risks associated to CTSC whole system. The current research made it possible to identify eight major groups of risks:

1. **Technical risks**

Technical issues associated to Capture, Transport and Storage are the ones which are developed in the previous pages.

2. **Risks related to CTSC project**

Mainly include the risks that affect the project progress, particularly the risks related to the project schedule, cost and performance; and development to commercial scales.

3. Social (Public acceptance) risks

Public acceptance is a risk that could significantly affect CTSC projects development. An example is Barendrecht project, in the Netherlands, which was cancelled due to public disagreement (CCJ 2010). De Coninck et al. believe that *the companies are not worried that CO_2 capture and storage will fail for technical reasons. One of the concerns, however, is potential public resistance to CCS, and some companies indicate that governments should step into provide neutral information to the lay public and it is imperative to find a common language for the characterization and communication of risk both among professionals and between professionals and the public* (De Coninck et al. 2009).

4. Policy/Strategy risks

Policy uncertainties are defined as a major risk to CTSC projects development. GCCSI defines four policy landscapes that affect CTSC technology (Fig. 1.5) (GCCSI 2011). CTSC is an innovative technology which is involved in global and local climate change and energy strategies. Therefore, the following policy issues could be concerned with CTSC.

Policies are not the same in different countries, and are strictly dependent of the policies regarding Climate Change.

5. Health, Safety and Environmental (HSE) risks

Technical matters, notably impurities, leakage and corrosion may lead to HSE problems. A number of HSE concerns have been already discussed in the previous sections.

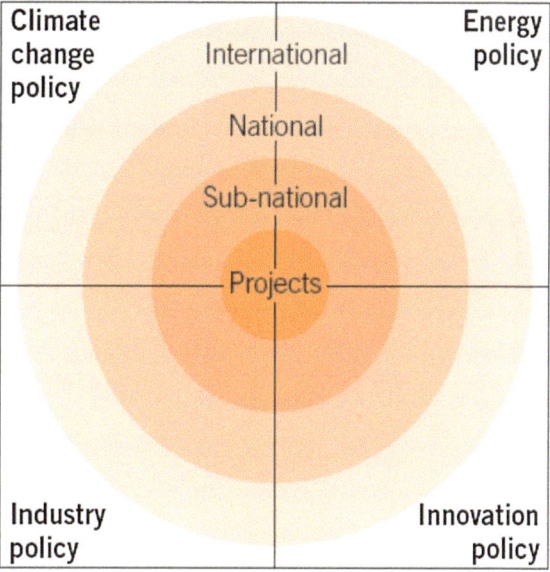

Fig. 1.5 Scope of policy landscapes related to CTSC (GCCSI 2011)

6. **Regulatory or legal risks**

According to a survey committed by GCCSI, regulatory issues are a significant challenge for CTSC projects (GCCSI 2011). Several international and regional regulations could cover the requirements of CTSC technology. These regulations need to be transposed into national or domestic laws. De Coninck et al. argue that IPPC Directive (96/61/EC, as amended) is applicable for CO_2 Capture in Europe. The IPPC Directive is the European Commission Directive on industrial emissions. The authors point out that *liquefied CO_2 is already transported in significant quantities by road, ship and pipeline across the EU and is regulated in accordance with dangerous goods laws and regulations.* However, Environmental Impact Assessment Directive (85/337/EEC, as amended) of European Commission could be applied for pipelines and pumping stations. EU Directive on CTSC *does not sufficiently deal with all legal uncertainties concerning the capture and transport of CO_2 derived from CCS facilities.* In spite of the availability of EU Directive for CTSC, *under current European law, it is uncertain whether CO_2 that is captured and then stored would be classified as 'waste'.* If so, the European waste laws could be applicable for CO_2 storage. This concern is currently the subject of several research studies (De Coninck et al. 2009).

7. **Organizational and human risks**

CTSC is a complex sociotechnical system which includes not only three technical components of Capture, Transport and Storage, but also an organizational structure containing a group of actors. The organizational and human risks are derived from such a complexity. The complex and sociotechnical systems will be defined later in this chapter.

8. **Financial/Economic risks**

As previously noted, some projects have been cancelled due to financial issues. GCCSI believes that project finance is the most challenging part of CTSC project successful development. Funding issues usually add uncertainties to the development of projects. Therefore, cost reduction is essential for the future of CTSC progress (Kapetaki and Scowcroft 2017).

Taking into account such an overview of risks, a list of thirty-nine risks is created based on several references, among others the documents of different projects such as Longannet, Lacq, Barendrecht, and the recent reports of GCCSI (GCCSI 2009, 2011; Longannet 2011; Feenstra et al. 2010; Kerlero de Rosbo 2009; CCP 2007; Lacq Project 2012; GCCSI 2016; Kapetaki and Scowcroft 2017).

The next step was to specify the project phase(s) related to each risk. The following six main phases are distinguished, which are not necessarily similar to GCCSI phases (presented in Fig. 1.3 and Table 1.1).

1. Opportunity:
   The beginning period, when negotiations are carried out on the feasibility of CTSC project.
2. Definition and planning:
   The phase when responsibilities and authorities of stakeholders are defined, and a planning is made for the project.
3. Engineering:
   Design and sizing of installations are performed in this phase.
4. Construction:
   This phase deals with construction and installation of required infrastructure and equipment.
5. Operation (Injection of CO_2):
   The period during which CO_2 is injected into the geological formation.
6. Post-injection (Monitoring) (also called "post-closure"):
   means the period after the closure of a storage site, including the period after the transfer of responsibility to the competent authority (EU Directive 2009).

Afterwards, the nature of each risk and the nature of consequences are identified. The risks are inevitably interconnected and could not be studied independently.

To analyze the reasons why a CTSC project does not progress as expected, the risks related to the very first phases of the project are extracted from the overall list. The result is a list of eighteen major risks (Table 1.4).

In Chap. 3, we will readdress these major risks and review the risks that could be analyzed with our systemic approach.

Table 1.4   Major risks affecting the very first phases of the project

Major risks affecting CTSC project progress (in the first phases)			
1	Project permits not obtained	10	Unavailability of a monetary mechanism for CO_2
2	Technology scale-up	11	Geographical infrastructure
3	Public opposition	12	Lack of financial resources
4	Lack of knowledge/qualified resources for operating the unit	13	Lack of political support
5	Legal uncertainties	14	High cost of project
6	Uncertainties in stakeholders' requirements/perceptions—communication problems	15	Unavailability of regulations regarding different types of storage (offshore/onshore)
7	Public availability of sensitive information	16	Uncertainties regarding the storage performance (capacity/injectivity/containment)
8	Change in policies/priorities	17	Model and data issues
9	Financial crisis impact on financial support of CCS projects	18	Uncertainties related to storage monitoring

1.4 Risk Management: Concepts and Evolution of Approaches

Before reviewing the evolution of risk management approaches, we need to introduce the definitions of RISK and RISK MANAGEMENT.

1.4.1 Definition of Main Concepts

1.4.1.1 Risk

Risk is often defined as a combination of two parameters: the probability and the severity of hazards.

From project management point of view, risk is an *"uncertain event or condition that, if it occurs, has a positive or negative effect on a project's objectives."* (PMBOK 2008).

The most comprehensive definition of risk in system safety engineering is the one specified by Leveson (1995). In her definition, risk is a combination of four components: hazard severity, hazard likelihood, hazard exposure and likelihood of hazard leading to an accident, as illustrated in Fig. 1.6.

1.4.1.2 Risk Management

Risk Management is defined in several references in different ways (Sadgrove 2005; Magne and Vasseur 2006; Desroches et al. 2006, 2007; Garlick 2007; Koivisto et al. 2009; Mazouni 2008). What will be referred as "Risk Management" in the present work is illustrated in Fig. 1.7.

Risk Management includes three main steps of analysis, evaluation and treatment of risk. In the risk analysis process, the scope is defined and the risks are identified and estimated. Afterwards, the risks are evaluated. The combination of risk evaluation and risk analysis is called risk assessment. Treatment is the final stage of risk management, where proposals for action are made and finally risks are reduced and controlled. The control process leads us to identify new risks or review

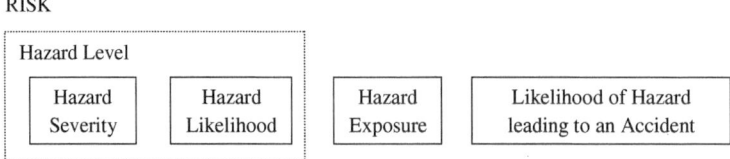

Fig. 1.6 Components of risk (Leveson 1995)

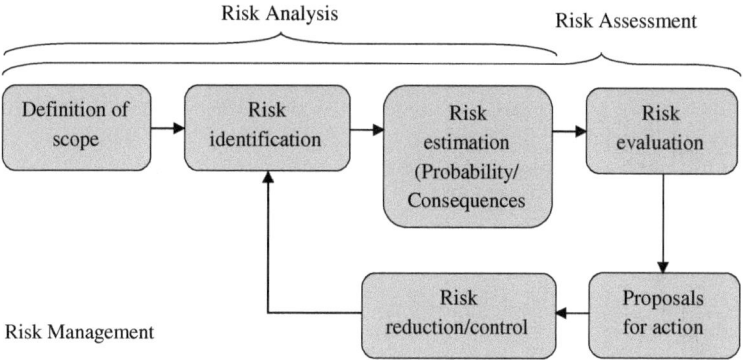

Fig. 1.7 Risk analysis, risk assessment and risk management process adapted from Koivisto et al. (2009) (based on IEC 60300-3-9: 1995, AS/NZS 4360: 2004, and ISO/IEC 73: 2002)

the previously defined ones, and go back to the risk analysis phase. This loop is shown in Fig. 1.7.

Risk acceptance and risk communication are specified as the last phases of risk management procedure by some authors (Condor et al. 2011).

1.4.2   Evolution of Risk/Safety Management Methods

Tixier et al. have already reviewed 62 risk analysis methodologies of industrial plants. The authors have categorized risk analysis methods in four main groups: deterministic, probabilistic, qualitative and quantitative. They conclude that a combination of several methods is necessary for making risk analysis more efficient.

Deterministic methods take into consideration the products, the equipment and the quantification of consequences for various targets such as people, environment and equipment. Probabilistic methods are based on the probability or frequency of hazardous situation apparitions or on the occurrence of potential accident. As they noted: The great majority of methods are deterministic, because historically operators and public organizations have initially tried to quantify damages and consequences of potential accidents, before to understand why and how they could occur (Tixier et al. 2002).

Quantitative Risk Assessment (QRA) methods aim to put figures on the likelihood and consequences of risk. A number of experts do not believe that quantitative approach is the best adapted way for modern complex sociotechnical systems. Some of them argue that semi-quantitative methods are less complicated and less time consuming (Dulac 2007; Kerlero de Rosbo 2009; Altenbach 1995).

According to Dulac, *to manage risk in complex engineering systems, it is necessary to understand how accidents happen* (Dulac 2007). Therefore, the notion of accident and accident models are introduced in this part.

Accident is an *unplanned and undesired loss event* which results in human, equipment, financial or information losses (Leveson 2009). Hollnagel defines "accident model" as a *stereotypical way of thinking about how an accident occurs* (Hollnagel 2004). Leveson (1995) believes that accident models can be used even for accident investigations or accident prediction.

Dulac describes that traditional risk analysis methods, such as Failure Modes and Effect Analysis (FMEA), Fault Tree Analysis (FTA) and Probabilistic Risk Assessment (PRA), are inappropriate for modern complex systems, because the interactions between different components of the system are not considered in these methods. The author also argues that organizational approaches *have made an important contribution to system safety by emphasizing the organizational aspects of accidents*. Even so, the organizational approaches often oversimplify the engineering part of the system (Dulac 2007). Hollnagel confirms the idea of Dulac, and proposes to find alternative methods of risk assessment for complex systems (Hollnagel 2004).

Based on this reasoning, Leveson has developed a new accident model, called STAMP (Systems-Theoretic Accident Model and Processes). This new accident model is based on systems theory concepts. In this sociotechnical model, she takes into account several actors of the system, from legislatures to company top management, project management, operations management and lower levels. She argues that lack of constraints imposed on the system design and on operations is the main cause of an accident, instead of a series of events. According to Leveson, STAMP model could be applied to any accidents in complex systems (Leveson 2004). The details of STAMP model approach will be presented in Chap. 2.

Another important notion is "systemic risk". Systemic risks are the *risks affecting the systems on which society depends. Systemic risks are characterized by complexity, uncertainty and ambiguity* (IRGC 2010; OECD 2003).

Hellström affirms that risks of innovative technologies are systemic, as they are connected to the social, economic and political infrastructure. The author argues that an integrated assessment of risk and innovation is indispensable. He believes that emerging technological innovation is systemic in the sense that it could not be separated from other aspects of the society. He suggests the integration of governance concepts in risk management, to make a systemic risk management approach (Hellström 2003). New risk management approaches are required for emerging technologies. The new approaches must involve all the stakeholders, including public. The public needs to be involved even in identification of risks. The author mentions that the new methodologies for analyzing emerging technological risk should be systemic (Hellström 2009).

In addition to integrity, the concept of "dynamic risk analysis" has been also remarked in the innovative risk analysis approaches. Garbolino et al. believe that *due to the complexity of the industrial systems and their own dynamic in time and space, the risk assessment methods need to be supported by a systemic vision of*

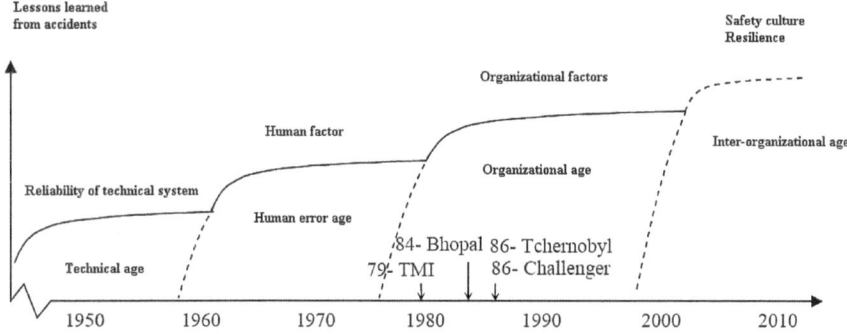

Fig. 1.8 Evolution of safety management approaches, translated from Cambon (2007) originally adapted from Groeneweg (2002), Wilpert and Fahlbruch (1998)

their processes. As they affirm in their article, modeling the industrial systems is indispensable to better understand their behavior in normal and abnormal modes (Garbolino et al. 2009).

Safety management approaches have evolved based on the lessons learned from industrial accidents (Fig. 1.8).

Until 1950s–1960s, safety was considered as a technical problem. Therefore, safety management was based on the improvement of technical systems reliability. From 1960s, technical issues were not sufficient to explain the accidents. Human errors are then brought in safety management approaches. By mid 1980s, lessons learned from industrial disasters such as Three Miles Island, Bhopal, Chernobyl and Challenger, highlighted the incomprehensiveness of human errors for analyzing accidents. According to Cambon, the lessons learned affirm that human errors could not be disconnected from the organizational context in which they had been generated. Hence, in 1980s–1990s, the human error is recognized as a consequence of the organizational problems. Cambon explains that the organizational approaches are characterized to be linear and epidemiological [as discussed by Hollnagel (2004)]. Systemic or inter-organizational age emerged at the beginning of the twenty-first century in order to answer to this weak point in the precedent (organizational) age (Cambon 2007).

Cambon intend to set off the significance of human and organizational factors in the management of safety within industries. However, it does not mean that technical issues have been completely removed from the causes of recent accidents since the technical age is terminated. This idea is supported by BARPI (Bureau d'Analyse des Risques et Pollutions Industriels). BARPI is the French office of Risk Analysis and Industrial Pollution, created in 1992, which is assigned to gather and analyze the information associated to industrial accidents (BARPI 2012).

In order to give a better structure to the Cambon's schematic (Fig. 1.8), and show the complementary evolution of the approaches, it is proposed to illustrate the evolution as shown in Fig. 1.9.

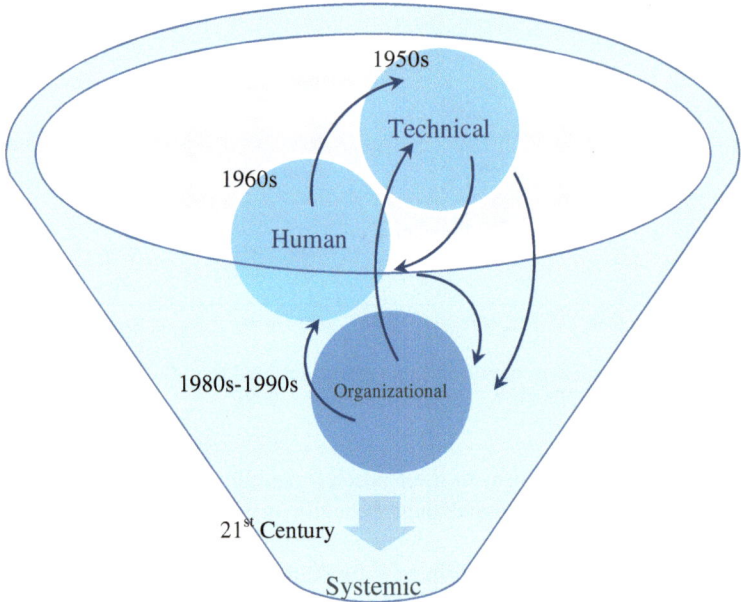

Fig. 1.9 Evolution of safety management approaches, an improved proposition

Figure 1.9 provides a clearer vision of the fact that Technical, Human and Organizational factors of safety are always in interconnection, and they cannot be disconnected through the evolution of approaches.

The funnel represents the systemic age, which includes all the three previous periods (organizational, human and technical). The arrows show the inter-relations of the three first ages.

1.5 Risk Management and CTSC

1.5.1 Available Risk Management Approaches for CTSC: Status and Limitations

So far, several works have been carried out on risk management of CTSC all around the world. In subsequent paragraphs, we will mention some examples of these methods to finally make out the necessity of developing a systemic framework for CTSC risk management.

Most of the available risk assessment or management methods are focused on one subsystem (Capture, Transport or Storage). Due to the uncertainties concerning the reliability of CO_2 storage, risk analysis studies particularly concentrate on the storage. A great number of studies only analyze the technical risks, and believe that the risks of Capture and Transport could be studied by classic methods.

1.5.1.1 CO_2 Capture: Available Risk Management Approaches

According to International Risk Governance Council (IRGC), risks associated to CO_2 capture technologies (except the innovative ones) are similar to a great number of industrial processes for which codes, standards and operating procedures have been developed. Consequently, risks related to CO_2 capture are currently well understood (IRGC 2009).

In France, analysis of major risks for CO_2 Capture is carried out based on ICPE (Bertrane 2011). ICPE (Installation Classée pour la Protection de l'Environnement) is the regulatory framework applicable to CO_2 Capture in France. ICPE is a French legislation for classified installations, transcribed from Seveso II. "Classified installation" is defined as *any industrial or agricultural operation likely to create risks or cause pollution or nuisance, notably in terms of local residents' health and safety* (ICPE website 1).

Seveso II is a European Directive on the control of major-accident hazards involving dangerous substances. The Directive *is aimed at the prevention of major accidents which involve dangerous substances, and the limitation of their consequences for man and the environment, with a view to ensuring high levels of protection throughout the Community in a consistent and effective manner* (Eurlex website).

Energy institute has recently published a guidance on hazard analysis of onshore CO_2 capture and pipeline structures. The major risk highlighted in this report is the risk of CO_2 leakage or energy release throughout the system, which may lead to equipment, human or economic losses. PHAST software is used to carry out the dispersion calculations. PHAST is a hazard analysis package developed by DNV (Energy Institute 2010).

1.5.1.2 CO_2 Transport: Available Risk Management Approaches

Neele et al. recommend that Quantitative Risk Assessment (QRA) methods, *as used for instance in the natural gas transportation industry*, could be used to study the HSE risks of CO_2 pipelines. In the report of CO_2Europipe project, the authors propose a standard risk management method for European pipeline infrastructure. DNV practice (DNV 2010) and ISO 31000 on Risk Management (ISO 31000 2010) are recommended to be used. *The aim of the CO_2Europipe project is to study the requirements for the development of a large-scale CO_2 transport infrastructure in Europe, between 2020 and 2050* (Neele et al. 2011).

The base of risk analysis for CO_2 transport in France is GESIP n°2008/01 (Safety Study guide, published by Groupe d'Etudes de Sécurité des Industries Pétroliers et chimiques) (Bertrane 2011).

Koornneef et al. from Utrecht University have reviewed the uncertainties regarding quantitative risk assessment of CO_2 transport by pipelines. They have studied the significant parameters in release and dispersion of CO_2 from pipelines and the effects on human beings' health. The assessed sources of uncertainties are:

failure rates, pipeline pressure and temperature, section length, diameter, orifice size, type and direction of release, meteorological conditions, jet diameter, vapor mass fraction in the release and the dose–effect relationship for CO_2 (Koornneef et al. 2010).

1.5.1.3 CO_2 Storage: Available Risk Management Approaches

EU Directive presents the risk assessment process of CO_2 storage in four steps: Hazard characterization, Exposure assessment, Effects assessment and Risk characterization. Definition of each step is summarized in the following paragraphs (EU Directive 2009):

1. Hazard characterization: means *characterizing the potential for leakage from the storage complex.* This includes specifying the potential leakage pathways, potential leakage rates, process specifications affecting potential leakage (e.g. maximum reservoir pressure, maximum injection rate and temperature).
2. Exposure assessment: is carried out *based on the characteristics of the environment and the distribution and activities of the human population above the storage complex, and the potential behavior and fate of leaking CO_2 from potential pathways.*
3. Effects assessment: includes the effects on *particular species, communities or habitats linked to potential leakage,* as well as the biosphere *(including soils, marine sediments and benthic waters).* The effect of CO_2 stream impurities and new substances generated through CO_2 storage shall be also studied.
4. Risk characterization: covers the safety and integrity aspects of the storage site in the short and long term. This step is performed based on the three previous steps of risk assessment, explained above.

Condor et al. have reviewed ten available risk assessment methodologies for CO_2 storage. The methods could be categorized in probabilistic/deterministic and qualitative/quantitative as previously noted from Tixier et al. (2002). The authors argue that quantitative methods are not appropriate for CO_2 storage at the current level of development, due to lack of required data. They believe that risks may be higher at the beginning of a CTSC project (Condor et al. 2011).

In France, BRGM is one of the predominant institutes working on risk management of CO_2 storage. BRGM specialists study the probable impacts of CO_2 leakage on drinkable water aquifers, human health and environment (Fabriol 2009; Bouc et al. 2009).

FEP (Features, Events, Processes) analysis is one of the approaches which has been already applied for risk assessment of CO_2 storage. FEP is a method for defining scenarios relevant to safety assessment of the geological disposal of radioactive wastes. The same approach has been implemented for long-term geological storage of CO_2. In the field of CO_2 storage, "Feature" refers to the geological formation and its characteristics. "Events" are what may or will happen in the future, for example earthquake. And "Processes" are the ongoing matters that influence the evolution of the system, like the erosion of the land surface. As a result of workshop discussions and brainstorming, a database has been developed for CO_2 storage including 200 FEPs in eight categories [for details, refer to Quintessa report (Savage et al. 2004)].

FEP analysis has been applied for different CO_2 storage projects, such as Weyburn (PTRC 2004) or Illinois Basin-Decateur project (Hnottavange-Telleen et al. 2009).

Oldenburg et al. have developed a certification framework to certify the effectiveness and safety of CO_2 storage. They have reviewed some available risk assessment methods for storage, including a system-modeling approach (CO_2 PENS), which studies the whole CTSC chain, from capture to storage. However, the authors believe that such comprehensive methods are so complex due to several uncertainties (Oldenburg et al. 2009).

Benson proposes to study lessons learned from analogous technologies in order to better understand the risks associated with CO_2 storage projects. She remarks three examples as the analogues of CO_2 storage: natural disasters like the catastrophic volcanic release of Lake Nyos in Cameroon, 1986, and the storage of natural gas and nuclear wastes (Benson 2002).

Perry has studied the experiences of natural gas storage industry and the potential application to CO_2 geological storage. He has reviewed the relevant literature and performed surveys/interviews with operators in Europe, Canada and the United States. *An important finding of this study is that only 10 of about 600 storage reservoirs operated in United States, Canada and Europe have been identified to have experienced leakage. Four due to cap rock issues, five due to well bore integrity, and one due to reservoir selection (too shallow).* Monitoring the geological formation is the most significant factor that he mentions for controlling the risks (Perry 2005).

1.5.1.4 CTSC Whole Chain: Available Risk Management Approaches

As discussed before, a great number of risk management approaches cover one aspect of CTSC chain. However, there are examples in the literature that highlight the necessity of an integrative risk management method for CTSC.

Farret et al. underline the importance of developing an integrated approach in risk analysis of CTSC due to interdependency of four steps, i.e. Capture, Transport, Injection to the reservoir and Long-term Storage (Farret et al. 2009).

Gerstenberger et al. believe that a comprehensive risk assessment method does not yet exist for CTSC and needs to be developed (Gerstenberger et al. 2009).

The (semi-)integrated studies that have been already carried out on CTSC risk management are as following:

GCCSI applies the Australian and New Zealand Standard for Risk Management (AS/NZS 4360: 2004) to define the likelihood and consequences of a set of extreme risks associated to integrated CTSC projects. Seventeen risks are identified with public, governmental/regulatory/policy, business case and technical nature (GCCSI 2009).

Det Norske Veritas (DNV) has studied HSE issues related to large-scale capture, transport and storage of CO_2. In their study, an almost integrated analysis has been performed; hence capture, transport and injection phases are considered in the analysis (storage phase is not included). DNV method for risk assessment of large-scale CTSC projects is SWIFT (Structured What IF Technique) analysis. SWIFT analysis is an expert panel/workshop approach to identify potential hazards and uncertainties. Prior to the workshop, a questionnaire was sent to the stakeholders in order to gather their ideas about HSE issues regarding CTSC (from capture to injection phase). The participants have mentioned the lack of an integrated approach as a concern in HSE risk management of CTSC, an approach that takes into account CTSC whole chain (Johnsen et al. 2009).

Another work on CTSC integrated risk analysis is the approach presented by Kerlero de Rosbo for the Belchatow project, in which Alstom was responsible to develop a CTSC plant for a coal-based power plant in Poland. Technical, financial, organizational, socio-political and regulatory risks associated with a large-scale CTSC project have been studied in that project. The deliverable was a risk register provided in panel discussions carried out to meet the project objectives (Kerlero de Rosbo 2009).

1.6 Requirement of a Novel Systemic Approach for CTSC Risk Management

CTSC is a complex sociotechnical system which includes a technical system with three components of Capture, Transport and Storage. The social part of CTSC sociotechnical system involves an organizational structure containing a group of actors. The interface between organizational, human and technical aspects could initiate a failure in the system. Here below, we will recall the definition of system, complex system, and sociotechnical system.

System
Durand points out six definitions for system (Durand 2010):

1. System is an organized whole, made up of interdependent elements that can be defined as relative to each other according to their place in this whole (definition of Ferdinand de Saussure, Swiss linguist)
2. System is a set of units and their mutual interrelations. (definition of Karl Ludwig von Bertalanffy, Austrian-born biologist)
3. System is a set of elements linked by a set of relationships (definition of Jacques Lesourne, French economist)
4. System is a set of elements in dynamic interaction which are organized based on a purpose. (Joël de Rosnay, French biologist)

5. System is a complex object, consisting of separate components interconnected by a number of relationships. (definition of Jean Ladrière, Belgian philosopher/logician)
6. System is a global unit organized by interrelationships between elements, actions or individuals. (definition of Edgar Morin, French philosopher and sociologist)

Complex System

International Risk Governance Council (IRGC) defines a complex system as a *system composed of many parts that interact with and adapt each other. In most cases, the behavior of such systems cannot be adequately understood by only studying their component parts. This is because the behavior of such systems arises through the interactions among those parts.* Complex systems have some common characteristics including *Emergence, Non-linearity, Inertia, Threshold behavior, and Hysteresis and Path Dependency.* These characteristics lead to difficulties in anticipating and controlling system behavior. IRGC argues that *Adaptability* and *Self-organization* are other features of complex systems that *make risk emergence less likely.* "Emerging risk" is defined as *one that is new, or a familiar risk that becomes apparent in new or unfamiliar conditions* (IRGC 2010).

Sociotechnical System

A sociotechnical system is a system consisting of a technical part that is in interaction with a social part. The components of sociotechnical system include human beings (workers, managers and all the stakeholders of internal and external environment), an organizational structure and a technical section (including equipment, methods and tools) (Carayon 2006). These components are in interrelation with the external environment of the system (Fig. 1.10) (Samadi and Garbolino 2011).

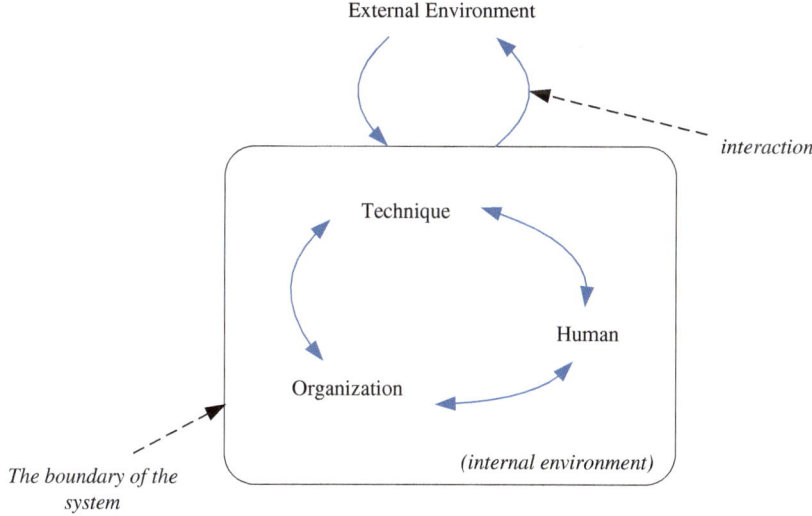

Fig. 1.10 Model of a sociotechnical system (Samadi and Garbolino 2011)

With the definitions provided in this section, we could consider CTSC as a complex sociotechnical system. As discussed before, traditional risk management methods are inappropriate for such systems, and novel systemic approaches are required.

A systemic approach will be presented in subsequent chapters for CTSC risk management. The proposed approach is based on systems theory concepts, system dynamics and STAMP, which will be introduced in Chap. 2.

References

Altenbach TJ (1995) A Comparison of risk assessment techniques from qualitative to quantitative. In: submitted to the ASME pressure and piping conference, Honolulu, Hawaii, 23–27 July 1995

BARPI (2012) Barpi: vingt ans d'accidents industriels, Bureau d'Analyse des Risques et Pollutions Industriels, Face au risque, n° 481, mars 2012. Accessed 23 Sept 2012. http://www.aria.developpement-durable.gouv.fr/ressources/em4_barpi_20_ans.pdf

Benson SM (2002) Comparative evaluation of risk assessment, management and mitigation approaches for deep geological storage of CO_2. Earth Sciences Division, E.O. Lawrence Berkeley National Laboratory

Benson SM (2007) Overview of geological storage of Co_2, Presentation in geological storage roundtable, 31 May 2007, Washington DC

Bertrane M (2011) L'évaluation des risques majeurs dans les projets de captage et transport de CO_2, Presentation of Centre Technique Industrie, Bureau Veritas, Presented in the international seminar on Capture, Transport and Storage of CO_2. Risk management: methods, practical experiences and perspectives, 7–8 Apr 2011, Le Havre, France

Bouc O, Audigane P, Bellenfant G, Fabriol H, Gastine M, Rohmer J, Seyedi D (2009) Determining safety criteria for CO_2 geological storage. Energy Procedia 1:2439–2446

BRGM (2005) Stockage Géologique du CO_2: analyses des risques, surveillance et mesures. Rapport Final, septembre 2005

Cambon J (2007) Vers une nouvelle méthodologie de mesure de la performance des systèmes de management de la sante-sécurité au travail, Thèse pour obtenir le grade de docteur de l'Ecole des Mines de Paris, Spécialité: Sciences et Génie des Activités à Risques, novembre 2007

Carayon P (2006) Human factors of complex sociotechnical systems. Appl Ergon 37(2006):525–535. https://doi.org/10.1016/j.apergo.2006.04.011

CCJ (2010) Shell Barendrecht project cancelled. Carbon Capture J, November 5, 2010. Accessed 22 June 2012. http://www.carboncapturejournal.com/displaynews.php?NewsID=676

CCP (2007) Public perception of carbon dioxide capture and storage: prioritized assessment of issues and concerns. Summary for policy-makers, CO_2 Capture Project®, March 2007. Accessed 20 July 2012. http://ccs101.ca/assets/Documents/iea_public_perception_of_ccs.pdf

Condor J, Unatrakarn D, Wilson M, Asghari K (2011) A comparative analysis of risk assessment methodologies for the geologic storage of carbon dioxide. Energy Procedia 4(2011):4036–4043. https://doi.org/10.1016/j.egypro.2011.02.345

De Coninck H, Flach T, Curnow P, Richardson P, Anderson J, Shackley S, Sigurthorsson G, Reiner D (2009) The acceptability of CO_2 capture and storage (CCS) in Europe: an assessment of the key determining factors, Part 1. Scientific, technical and economic dimensions. Int J Greenhouse Gas Control 3(2009):333–343. https://doi.org/10.1016/j.ijggc.2008.07.009

Desroches A, Leroy A, Vallée F (2007) La gestion des risques, Lavoisier, p 101

Desroches A, Leroy A, Quaranta JF, Vallée F (2006) Dictionnaire d'analyse et de gestion des risques, Lavoisier, p 23

De Visser E, Hendriks Ch, Barrio M, Mølnvik M, De Koeijer G, Liljemark S, Le Gallo Y (2008) Dynamis CO_2 quality recommendations. Int J Greenhouse Gas Control 2(2008):478–484. https://doi.org/10.1016/j.ijggc.2008.04.006

DNV (2010) Design and operation of CO_2 pipelines Recommended practice DNV-RP-J202. Det Norske Veritas, Norway, Apr 2010

Dulac N (2007) A framework of dynamic safety and risk management modeling in complex engineering systems. PhD thesis submitted to the department of aeronautical and astronautical engineering at MIT, Feb 2007

Durand D (2010) La systémique, Presses Universitaires de France (PUF), 11th edn, Jan 2010

Energy Institute (2010) Technical guidance on hazard analysis for onshore carbon installations and onshore pipelines, published by Energy Institute, London, 1st edn, Sept 2010

EPA (2016) United States Environmental Protection Agency. Global Greenhouse Gas Emissions Data. https://www3.epa.gov/climatechange/ghgemissions/global.html#three. Accessed 26 Mar 2016

ESRL (2017) Earth system research laboratory global monitoring division. Trends in Atmospheric Carbon Dioxide. https://www.esrl.noaa.gov/gmd/ccgg/trends/global.html. Accessed 30 July 2017

EU Directive (2009) Directive 2009/31/EC of the European Parliament and of the Council of 23 April 2009, on the geological storage of carbon dioxide and amending Council Directive 85/337/EEC, European Parliament and Council Directives 2000/60/EC, 2001/80/EC, 2004/35/EC, 2006/12/EC, 2008/1/EC and Regulation (EC) No 1013/2006

Eurlex website, Official Journal of European Union, Official Journal L 010, 14/01/1997, 0013–0033. Accessed 12 June 2012. http://eur-lex.europa.eu/LexUriServ/LexUriServ.do?uri= CELEX:31996L0082:EN:HTML

Fabriol H (2009) Risk management and uncertainties: ensuring CO_2 storage safety. In: 3rd international symposium: capture and geological storage of CO_2, 5–6 Nov 2009, Paris

Farret R, Gombert P, Lahaie F, Roux P (2009) Vers une méthode d'analyse des risques globale de la filière CSC, intégrant plusieurs échelles de temps, Rapport Scientifique, INERIS, 2008-2009, pp 100–110

Feenstra CFJ, Mikunda T, Brunsting S (2010) What happened in Barendrecht? Case study on the planned onshore carbon dioxide storage in Barendrecht, the Netherlands, ECN/CAESAR, 3 Nov 2010

Garbolino E, Chéry JP, Guarnieri F (2009) Dynamic systems modelling to improve risk analysis in the context of Seveso industries. Chem Eng Trans 17:373–378

Garlick A (2007) Estimating risk. A management approach. Gower Publishing Limited, England, pp 9–31

GCCSI (2009) Strategic analysis of the global status of carbon capture and storage. Report 5: synthesis report, Global CCS Institute, Canberra, Australia

GCCSI (2011) The global status of CCS: 2011. Global CCS Institute 2011, Canberra, Australia. ISBN 978-0-9871863-0-0

GCCSI (2012) The global status of CCS: 2012. Global CCS Institute 2012, Canberra, Australia. ISBN 978-0-9871863-1-7

GCCSI (2016) The global status of CCS: 2016. Global CCS Institute 2016, Summary Report, Australia. ISBN 978-0-9944115-6-3

GCCSI (2017a) Large-scale CCS facilities Project Database. https://www.globalccsinstitute.com/projects/large-scale-ccs-projects. Accessed 6 Oct 2017

GCCSI (2017b) Large-scale CCS facilities Definition. https://www.globalccsinstitute.com/projects/large-scale-ccs-projects-definitions. Accessed 6 Oct 2017

Gerstenberger M, Nicol A, Stenhouse M, Berryman K, Stirling M, Webb T, Smith W (2009) Modularised logic tree risk assessment method for carbon capture and storage projects. Energy Procedia 1:2495–2502

Groeneweg J (2002) Controlling the controllable, preventing business upsets, 5th edn. Global Safety Group, Den Haag

Hellström T (2003) Systemic innovation and risk: technology assessment and the challenge of responsible innovation. Technol Soc 25(2003):369–384. https://doi.org/10.1016/S0160-791X(03)00041-1

Hellström T (2009) New vistas for technology and risk assessment? The OECD programme on emerging systemic risks and beyond. Technol Soc 31(2009):325–331. https://doi.org/10.1016/j.techsoc.2009.06.002

Hnottavange-Telleen K, Krapac I, Vivalda C (2009) Illinois Basin-Decateur Project: initial risk assessment results and framework for evaluating site performance. Energy Procedia 1 (2009):2431–2438

Hollnagel E (2004) Barriers and accident prevention. Published by Ashgate publishing company, England. ISBN 0754643018

ICPE website 1. http://www.installationsclassees.developpementdurable.gouv.fr/Definition,14731.html. Accessed 12 June 2012

IEA (2017) Energy technology perspectives 2017. International Energy Agency, OECD/IEA, Paris, France

IPCC (2005) Special report on carbon dioxide capture and storage, Intergovernmental Panel on Climate Change. Cambridge University Press, London

IRGC (2009) Power plant CO_2 capture technologies, risks and risk governance deficits. International Risk Governance Council, Geneva. ISBN 978-2-9700672-2-1

IRGC (2010) The emergence of risks: contributing factors. Report of International Risk Governance Council, Geneva, 2010. ISBN 978-2-9700672-7-6

ISO 31000 (2010) Risk management. Principles and guidelines, Jan 2010. NF ISO 31000:2010-01

Johnsen K, Holt H, Helle K, Sollie OK (2009) Mapping of potential HSE issues related to large-scale capture, transport and storage of CO_2. DNV Report for Petroleumstilsynet

Kapetaki Z, Scowcroft J (2017) Overview of Carbon Capture and Storage (CCS) demonstration project business models: risks and enablers on the two sides of the Atlantic. Energy Procedia 114(2017):6623–6630. https://doi.org/10.1016/j.egypro.2017.03.1816

Kerlero de Rosbo G (2009) Integrated risk analysis for large-scale CCS projects implementation. Thèse professionnelle du Master Spécialisé en Ingénierie et gestion de l'Environnement (ISIGE)

Koivisto R, Wessberga N, Eerolaa A, Ahlqvista T, Kivisaaria S, Myllyojaa J, Halonena M (2009) Integrating future-oriented technology analysis and risk assessment methodologies. Technol Forecast Soc Change 76:1163–1176

Koornneef J, Spruijt M, Molag M, Ramirez A, Turkenburg W, Faaij A (2010) Quantitative risk assessment of CO_2 transport by pipelines—a review of uncertainties and their impacts. J Hazard Mater 177:12–27

Lacq Project (2012), Documents concerning Lacq Project, available on the regional prefecture website. Accessed from Nov 2010 to Oct 2012. http://www.pyrenees-atlantiques.pref.gouv.fr/Politiques-publiques/Environnement-risques-naturels-et-technologiques/Pilote-d-injection-de-CO2/Projet-de-captage-CO2-de-Total/

Leveson N (1995) Safeware, system safety and computers. Addison-Wesley Publishing Company. ISBN 0201119722

Leveson N (2004) A new accident model for engineering safer systems. Saf Sci 42:237–270

Leveson N (2009) Engineering a safer world, system safety for the 21st century, Massachusetts Institute of Technology, July 2009 (a draft book). Accessed 1st Dec 2011. http://sunnyday.mit.edu/safer-world.pdf

Longannet (2011) FEED documents of Longannet project, published online http://www.decc.gov.uk/en/content/cms/emissions/ccs/demo_prog/feed/scottish_power/scottish_power.aspx. Accessed 29 Nov 2011

Magne L, Vasseur D (2006) Risques Industriels, complexité, incertitude, et décision: une approche interdisciplinaire, Lavoisier, collection EDF R&D, p 181

Mazouni MH (2008) Pour une Meilleure Approche du Management des Risques: De la Modélisation Ontologique du Processus Accidentel au Système Interactif d'Aide à la Décision, Thèse, Doctorat de l'Institut National Polytechnique de Lorraine, Spécialité: Automatique, Traitement du Signal et Génie Informatique, Novembre 2008, pp 39–42

MIT (2016) Massachusetts Institute of Technology, Cancelled or Inactive Projects. http://
    sequestration.mit.edu/tools/projects/index_cancelled.html. Accessed 4 July 2017
Neele F, Mikunda T, Seebregts A, Santen S., Van der Burgt A, Nestaas O, Apeland S, Stiff S,
    Hustad C (2011) Developing a European CO_2 transport infrastructure, WP 1.1 Needs for a
    future framework for CCS, Revision 19, TNO. Accessed 12 June 2012. http://www.
    co2europipe.eu/Publications/CO2Europipe%20-%20Executive%20Summary.pdf
OECD (2003) Emerging risks in the 21st century. An agenda for action, organization for economic
    co-operation and development. OECD Publications Service, France
Oldenburg CM, Bryant SL, Nicot JPh (2009) Certification framework based on effective trapping
    for geologic carbon sequestration. Int J Greenhouse Gas Control 3:444–457. https://doi.org/10.
    1016/j.ijggc.2009.02.009
Perry KF (2005) Natural gas storage industry experience and technology: potential application to
    CO_2 geological storage. In: Thomas DC, Benson SM (eds) Carbon dioxide capture for storage
    in deep geological formations. Elsevier Ltd.
PMBOK (2008) A guide to the project management body of knowledge (PMBOK® Guide), 4th
    edn, Project Management Institute, USA. ISBN 978-1-933890-5-7
PTRC (2004) IEA GHG Weyburn CO_2 monitoring & storage project: summary report 2000-2004.
    In: from the proceedings of the 7th international conference on greenhouse gas control
    technologies, 5–9 Sept 2004, Vancouver, Canada
Rochon E, Kuper J, Bjureby E, Johnston P, Oakley R, Santillo D, Schulz N, Von Goerne G (2008)
    False hope: why carbon capture and storage won't save the climate, Greenpeace, Published in
    May 2008 by Greenpeace International, Amsterdam, The Netherlands
Sadgrove K (2005) The complete guide to business risk management, 2nd edn. Gower Publishing
    Limited, England, pp 277–284
Samadi J, Garbolino E (2011) A new dynamic risk analysis framework for CO_2 capture, transport
    and storage chain. In: 29th international conference of system dynamics society, 24–28 July
    2011, Washington DC
Savage D, Maul PhR, Benbow S, Walke RC (2004) A generic FEP database for the assessment of
    long-term performance and safety of the geological storage of CO_2. Quintessa report, June
    2004
Seiersten M (2001) Material selection for separation, transportation and disposal of CO_2. In:
    Proceedings of the corrosion 2001. National Association of Corrosion Engineers, paper 01042.
    Accessed 15 June 2012. http://www.onepetro.org/mslib/app/Preview.do?paperNumber=
    NACE-01042&societyCode=NACE
Serpa J, Morbee J, Tzimas E (2011) Technical and economic characteristics of a CO_2 transmission
    pipeline infrastructure. JRC report, European Union, EUR 24731 EN. ISBN
    978-92-79-19425-2, ISSN 1018-5593. Accessed 17 Mar 2011. https://doi.org/10.2790/30861,
    http://publications.jrc.ec.europa.eu/repository/bitstream/111111111/16038/1/reqno_jrc62502_
    aspublished.pdf
Tixier J, Dusserre G, Salvi O, Gaston D (2002) Review of 62 risk analysis methodologies of
    industrial plants. J Loss Prev Process Ind 15:291–303
UNDP (2007) Human Development report 2007/2008, Fighting climate change: human solidarity
    in a divided world, Published for the United Nations Development Programme (UNDP), USA.
    ISBN 978-0-230-54704-9
Wilpert B, Fahlbruch B (1998) Safety related interventions in inter-organisational fields, Chapter
    14, pp 235–248. In Hale & Baram, 1998, Safety Management, The challenge of change,
    Pergamon
Wright I (2011) Geologic CO_2 storage assurance at In Salah, JIP phase 1: storage capacity
    assessment. Presented in: CSLF projects interactive workshop. 1–2 March 2011, Al Khobar,
    Saudi Arabia, Accessed 18 June 2012. www.cslforum.org/publications/documents/
    alkhobar2011/GeologicCO2StorageAssurance_InSalahAlgeria_Session1.pdf

Chapter 2
Systems Theory, System Dynamics and Their Contribution to CTSC Risk Management

Abstract This chapter deals with the concepts of systems theory, system dynamics and STAMP (Systems-Theoretic Accident Model and Processes). Feedback, Feedback loop, Causal graph and Delay are introduced as significant notions of system dynamics, which are essential to go on to the next chapter. "Endogenous point of view" is presented as a foundation of system dynamics, which provides endogenous explanations for all phenomena. A main question about CTSC at the present time is whether the technology will be developed progressively up to commercial scales. The question is reformulated in a systems thinking framework to study how dynamics of risks can affect dynamics of CTSC projects, and how interconnections of stakeholders and associated risks can result in the success or failure of a CTSC project. Current dynamics of CTSC are reviewed for the purpose of formulating this question. STAMP major concepts, including safety constraint, hierarchical control structure and process model, are presented in order to understand how STAMP can contribute to study the risks of CTSC projects.

In this chapter, we will argue how systems theory, system dynamics, and STAMP (Systems-Theoretic Accident Model and Processes) approach can contribute to the risk management of CTSC projects. The chapter includes three sections:

The first section is devoted to the introduction of systems theory and system dynamics.

In the second section, current dynamics of CTSC technology are presented.

We will finally discuss how systemic approaches, and specially STAMP, could analyze the dynamics of CTSC.

© The Author(s) 2018   27
J. Samadi and E. Garbolino, *Future of CO₂ Capture, Transport and Storage Projects*,
SpringerBriefs in Environmental Science, https://doi.org/10.1007/978-3-319-74850-4_2

2.1 Systems Theory and System Dynamics: Introduction and Key Concepts

2.1.1 Systems Theory

Development of the modern systems theory was essentially localized in the United States. Nevertheless, in 1960s, 1970s the phenomenon was introduced out of the US, essentially by two publications: "The limits to Growth", 1972 (Meadows et al. 1972) and "Le macroscope", published in France in 1975 (De Rosnay 1975).

Prior to the emergence of systemic thinking, occidental science was built on "classic rationalism" of Aristotle and Descartes.

There are four principal concepts in the systemic approach: interaction, totality, organization and complexity (Durand 2010):

- **Interaction**: means the mutual effect of system elements, comparing to the simple causal action of A on B in classic science. This notion leads us to the concept of feedback, which will be defined later.
- **Totality**: the best and most ancient citation regarding this concept is the Blaise Pascal's (French mathematician, physicist, inventor, philosopher, religious thinker, and writer of the 17th century). He believes that it is impossible to know the parts without knowing the whole and to know the whole without knowing the individual parts.
- **Organization**: could be defined as a set of relations among the elements or individuals which forms a new unit, without necessarily the same qualities of the components.
- **Complexity**: Edgar Morin is one of the pioneers in complexity. He believes that complexity refers not only to the quantity of elements and interactions in the system, but also to the uncertainties, disinclinations and unpredictable phenomena (the concept of "emergence" in complex systems which was introduced in Chap. 1). Morin states that complexity always deals with hazard (Morin 2005).

2.1.2 System Dynamics

System dynamics has its roots in control engineering, cybernetics and general system science (Fuchs 2006).

System dynamics is a methodology to understand the structure and behavior of complex systems, created during the mid 1950s by Jay W. Forrester in the Massachusetts Institute of Technology (MIT). He defines system dynamics as a combination of theory, methods and philosophy required to analyze the behavior of systems (Forrester 1991).

System dynamics is grounded in the theory of nonlinear dynamics and feedback control developed in mathematics, physics and engineering (Sterman 2000).

So far, system dynamics has been applied in various fields from management to environmental change, politics, economic behavior, medicine, engineering, and recently for analyzing accidents and risks (Forrester 1991; Leveson 2004a, b; Stringfellow 2010; Garbolino et al. 2009, 2010; Dulac 2007).

Models and modeling are the most essential concepts in system dynamics. Models are simplifications of reality which help people to clarify their thinking and improve their understanding of the world. Paul Valéry (French writer, poet, philosopher and epistemologist) believes that models are the only bases of our thinking (Durand 2010). *All human beings have a mental representation of the systems around them, such as families, universities, cities, etc. These mental models are flexible and rich in detail, but they are often fuzzy, incomplete and imprecise. That's why system dynamicists propose to decision-makers to apply system dynamics to map out their mental models on the computer and follow the evolution of the system through the computer model* (Radzicki and Taylor 1997). The process of modeling a system and studying its behavior over time is termed "dynamic modeling".

System dynamics modeling consists of four main concepts of Variable, Feedback (loop), Causal Graph and Delay. These notions are introduced in the subsequent paragraphs.

- **Variable**: Each element of the system that we put in the model could be a variable.

  There are three types of variables: Stock, Flow, and Auxiliary (Control) variables.

  - **Stock (or Level)**: is an accumulation within the system, for example inventories, goods in transit, bank balances, factory space and the number of employees (Forrester 1968). The quantity of stock is the integral of difference between its outflow and inflow.
  - **Flow (or Flow Rate or Rate)**: is another element of a model structure that defines the flows between the stocks in the system (Forrester 1968).

    One of the most challenging parts of dynamic modeling is to correctly distinguish the stocks and the flows. As Forrester proposes, in order to determine whether a variable is a stock or a flow, we should see whether or not the variable would continue to exist in the system. The variable which will continue to exist is a stock (such as inventories of a warehouse). Flow is the variable that could be stopped (like receiving and shipping goods).
  - **Auxiliary (Control) variable**: is a variable that *is computed from other variables at a given time. Auxiliaries are typically the most numerous variable type* (Vensim 2010).

- **Feedback and Feedback Loop**: *From a system dynamics point of view, a system can be classified as "open" or "closed".* Open systems are the ones in which the outputs have no influence on the inputs of the system. In closed systems, the outputs do have influence on the inputs. Most of the systems in the real world are closed systems and the effect of output on input is called Feedback (Radzicki and Taylor 1997).

  There are two types of feedback loops in a closed system: positive (or reinforcing) and negative (or balancing) loops. Positive feedback loops are the ones which *destabilize the system and cause them to run away from their current situation* whilst negative feedback loops stabilize the system (Radzicki and Taylor 1997). In a positive feedback loop, the variables change in the same direction, whereas they will change in the opposite direction in a negative feedback loop. In other words, in a positive feedback, if the first variable increases, the second variable will be increased (the same direction as the first variable). However, in a negative feedback, an increase in the first variable leads to a decrease in the second variable (the opposite direction).

  Negative and positive feedback loops in a complex system result in the synchronization of the system and help the system to keep its dynamic equilibrium. Feedback loops depend on a series of decisions from different endogenous and exogenous actors (Louisot 2004).

  In order to better understand the concept of feedback, we could review an example (Fig. 2.1). If we consider the balance between gasoline consumption and car pools, in case of an increase in gasoline consumption, gasoline price will increase. This is a positive feedback as shown in Fig. 2.1.

  The increase of gasoline price will motivate people to join car pools. As a result, the number of vehicles and gasoline consumption will be reduced. This is a negative feedback loop which is illustrated in Fig. 2.1.

- **Causal graphs**: are the diagrams containing the network of feedback loops, or the interactions of the system variables.

  The examples concerning CTSC system will be presented in Chap. 3.

- **Delay**: is another important concept in system dynamics. Radzicki and Taylor explain the concept of delay by the fact that *events in the world do not occur instantaneously. Instead, there is often a significant lag between cause and effect. The longer the delay between cause and effect, the more likely it is that a decision maker will not perceive a connection between the two* (Radzicki and Taylor 1997).

Fig. 2.1 An example of negative feedback loop (Radzicki and Taylor 1997)

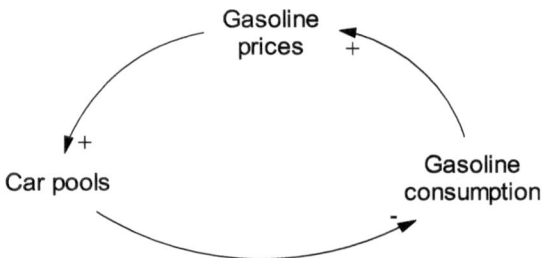

Several software packages are available for dynamic modeling such as VENSIM®, STELLA®, GOLDSIM®, ITHINK® and POWERSIM®. Models presented in the current work have been developed by VENSIM®.

Richardson (2011) believes that "endogenous point of view" is in fact the foundation of system dynamics. In this viewpoint, system is considered as cause. *System dynamicists use system thinking, management insights and computer simulation to*:

- *hypothesize, test and refine endogenous explanations of system change, and*
- *use those explanations to guide policy and decision making* (Richardson 2011).

Sterman asserts Richardson idea and mentions that *system dynamics seeks endogenous explanations for phenomena. "Endogenous"* is defined as *"arising from within"* by Sterman (2000).

2.2 Current Dynamics of CTSC

Dynamic complexity arises because of certain characteristics of systems, and among others because systems are dynamic, tightly coupled, governed by feedback, nonlinear, history-dependent and self-organizing (Sterman 2000).

Several sorts of dynamics are involved in CTSC current context. The main categories of dynamics are as following:

2.2.1 Dynamics of Climate/Atmosphere

The temperature of the earth's surface is increasing, mainly because of anthropogenic greenhouse emissions, which have been growing exponentially since the beginning of the industrial age. Greenhouse gases, such as CO_2, absorb a part of the energy radiated by the earth. Therefore, the amount of energy radiated back by the earth into the atmosphere will be less than the insolation. Consequently, the earth's surface temperature increases (Sterman and Sweeney 2002).

Nevertheless, there are controversial ideas on the sources of global warming. Some have an endogenous view, and believe that human activities are responsible for global warming. From the contrary exogenous point of view, the increase of CO_2 concentrations and global temperature is part of a natural phenomenon (Richardson 2011). The endogenous viewpoint on the climate dynamics explains the necessity to mitigate industrial CO_2 emissions. CTSC is one of the mitigation options.

Dynamics of global temperature rise and atmospheric CO_2 concentrations are presented in Fig. 2.2.

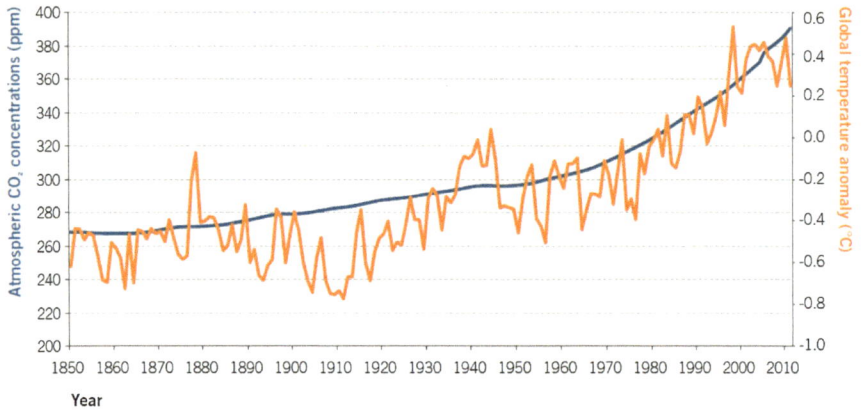

Fig. 2.2 Global CO_2 atmospheric concentrations and temperature (GCCSI 2011)

2.2.2 Dynamics of Subsurface

A dynamic system is defined as a system which is in motion, when material and energy change from one form to another.

There are two types of systems in geology:

– A closed system that exchanges only heat (no matter) with its environment
– An open system that exchanges both heat and matter with its surroundings

Most geological systems are open systems, in which matter and energy freely flow across the system's boundaries. Therefore, materials on and in earth are changed and rearranged. The direction of change in a dynamic geological system, and generally in natural systems is towards a state of equilibrium. *The equilibrium is a condition of the lowest possible energy, or a condition in which the net result of the forces acting on a system is zero* (Hamblin and Christiansen 2004).

The geological environment, where CO_2 is injected and stored, is dynamic. The variations are under control by different modeling tools. The purpose is to make sure that the injected CO_2 will be remained isolated from the other compartments of the geological formation above the caprock (low-permeable geological layer that assures the sealing of CO_2 injection reservoir).

2.2.3 Dynamics of Project

Project is the third aspect for which dynamics could be studied. CTSC projects have some common points with other industrial projects. There are also some specific characteristics, since CTSC is a novel technology and several actors and stakeholders are engaged in the development process of the technology.

Typical project dynamics, which are subject of several studies, include project staffing and productivity (Lyneis et al. 2001). Stakeholders and project phases dynamics are other aspects of project dynamics. Stakeholder dynamics is defined as the potential complex behavior of stakeholders interacting over time. Interactions of stakeholders with different goals and perceptions of the system generate essential feedback effects within the system (Richardson and Andersen 2010). Several stakeholders are involved in CTSC projects. Governments (national and local), project developers, local public, municipal and regulatory authorities, and non-governmental organizations (such as environmental organizations) are the main stakeholders of a CTSC project. The second aspect of project dynamics is related to the project phases. The major phases of a CTSC project consist of Opportunity, Definition and Planning, Engineering, Construction, Operation (Injection of CO_2), and Post-injection (Monitoring). These phases have been already defined in Chap. 1.

2.2.4   Dynamics of Risks

Emerging risks are dynamic, since the systems are regularly adapting themselves to perturbations. *Some emerging risks lessen over time while others become worse than anticipated.* Therefore, the consequences of emerging risks are not easily predictable. Furthermore, time delays between the perturbations, system responses and the internal/external impacts complicate the identification of emerging risks.

"Positive feedback" is one of the factors that could lead to the emergence of systemic risk. When the system response to a perturbation creates amplifications and destabilizes the system, a positive feedback is present. The notion of feedback (positive and negative) has been explained earlier in this chapter. Positive feedbacks tend to be destabilizing. Hence, they can *potentially increase the likelihood or consequences of the emergence of a new, systemic risk. It is therefore important for analysts to identify feedback dynamics (both positive and negative) that are occurring in a system, and assess their function and their relative balance (if their positive or negative dominate) in order to better anticipate when risks might emerge or be amplified* (IRGC 2010).

Rodrigues believes that risks are dynamic events. Risk dynamics are generated by a network of feedback loops in the project. He affirms that the management needs to have a systemic view to understand why risks emerge, because risks have a systemic nature (Rodrigues 2001).

With this introduction of CTSC current dynamics, we can now move forward to discuss how these dynamics could be studied and analyzed using systemic approaches.

2.3 Contribution of Systemic Approaches and System Dynamics to Study the Dynamics of CTSC

CTSC is a complex sociotechnical system, including three technical components of Capture, Transport and Storage, and an organization structure containing a group of actors (Samadi and Garbolino 2011).

Available lessons learned from CTSC projects confirm that the feedback loops of different types of risk are significant in the development process of projects. Technical aspects of long-term safety of CO_2 storage have been always at the heart of risk assessment studies. However, technical risks are continually in inter-relation with other aspects of risk.

At the present time, the main question about CTSC is whether the technology will be developed progressively up to commercial scales, as many projects have been cancelled during the previous years.

In the current work, we focus on modeling risks of CTSC projects development using a systemic approach. In other words, dynamics of project and risks are under study.

The question is how dynamics of risks affect dynamics of CTSC projects, and how interconnections of stakeholders and associated risks could result in the success or failure of a CTSC project progress. If we rephrase the goal in system dynamics language, the problem we are modeling is that some particular CTSC projects are not successful in the development.

A methodology is developed based on STAMP approach to model the structure of safety control in CTSC projects, and analyze the feedback network dynamics of CTSC project risks.

STAMP approach has its roots in the control theory. Leveson presents three main concepts in STAMP model. These concepts are: safety constraints, hierarchical control structures and process models.

Safety constraint is a major notion in STAMP. Leveson argues that *events leading to losses only occur because safety constraints were not successfully enforced. Therefore, we first need to identify the safety constraints to enforce and then to design effective controls to enforce them* (Leveson 2009).

Hierarchical control structures are the basis of systems in systems theory. Mutual feedbacks of controllers (each level of hierarchical control structure) lead to the improvement of maintaining safety constraints. Leveson proposes a general model of sociotechnical system control based on the model previously presented by Rasmussen. Leveson model is illustrated in Fig. 2.3.

Figure 2.3 shows the hierarchical control structure of a system in two phases of development and operations. Documents, procedures and policies exchanged between different levels of the structure are demonstrated on the arrows.

Process models are the third significant notion in STAMP. Leveson remarks that *any automated or human control needs a model of the process being controlled to control it effectively.* Process models must include the *relationships among the*

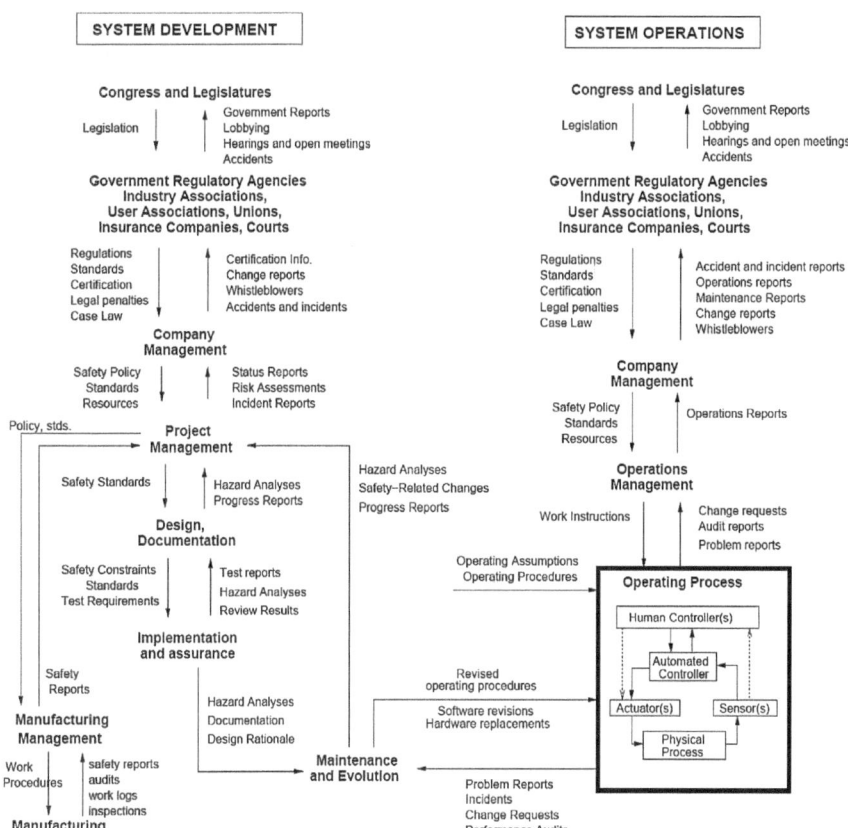

Fig. 2.3 Model of sociotechnical system control structure (Leveson 2009)

system variables, the current state, and the ways the process can change state (Leveson 2009).

She believes that in order to understand why accidents happen in a system and to prevent losses in future, we first need to review the control actions already available in the system. These control actions could be translated as safety constraints. Then we should review why and how inadequate control actions will lead the system to a hazardous state.

STAMP model could be applied either for analyzing accidents which have already happened or for evaluating the safety in a system, where an accident has not occurred yet.

Control system engineering emerged in 1930s, when engineers began building automatic control systems by using the techniques of electronics (Powers 1990). Leveson schematizes a standard control loop as shown in Fig. 2.4:

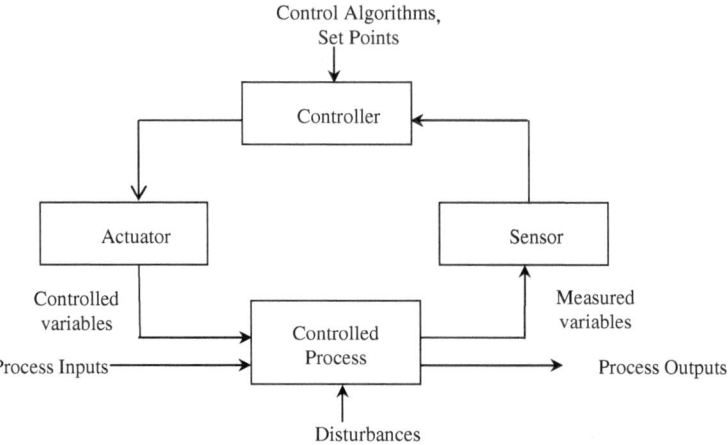

Fig. 2.4 A standard control loop (Leveson 2009)

To control a process, the controller must have four conditions:

1. Having a goal
2. Be able to affect the system
3. Be or contain a model of the system
4. Be able to observe the system

The controller observes the system by sensors, which obtains the measured variables of the system. The output of the controller affects the system by providing controlled variables through actuators. The purpose is to maintain the set point, which is the goal of the controller (Leveson 2009).

In the field of our study, the process under control is the progress of CTSC project in a sustainable manner. The actors or stakeholders of CTSC technology are the controllers.

This idea will be explained in Chap. 3.

Leveson explains the steps of applying STAMP model to analyze accidents as follows (Leveson 2009):

1. First of all, we should find out the events leading to the loss. It means that we simply list all the chain of events contributing in the occurrence of that accident.
2. Secondly, the hazards and system boundaries should be identified.
3. The next step is to find out the system safety constraints and system requirements regarding each hazard.
4. Then, we should form the hierarchical structure of safety control for the system. The roles and responsibilities of each actor (controller) should be clearly defined in this structure.
5. After that, the losses of physical system level should be analyzed. Four principal categories of information are required in this analysis including: safety

requirements and constraints, existing controls, failures and inadequate controls, and the context.

6. The sixth step is to analyze the hierarchical levels of safety control structure. In this stage, we need to collect four groups of information for each actor: safety-related responsibilities, context, unsafe decisions and control actions, and process model flaws.

7. In the next step, the coordination and communication between actors (controllers) will be studied.

8. Subsequently, we should study the dynamics (changes over time) relating to the loss.

9. The final step is to offer recommendations to prevent similar accidents in future.

A systems-theoretic hazard analysis method is also developed by Leveson. This method, called STPA (Systems-Theoretic Process Analysis), is applied for the assessment of safety in a system, when an accident has not happened yet.

The obvious difference between this case and the case when we analyze an accident by STAMP is that in the former we cannot definitely identify the inadequate control actions. Alternatively, we could analyze the "potential" inadequate or insufficient safety constraints.

The steps of STPA analysis could be summarized as following (Pereira et al. 2006):

1. Review the hazards and ensure that safety constraints are in place
2. Model the hierarchical structure of safety control in the system
3. Identify potentially inadequate control actions
4. Determine how potentially inadequate control actions could lead to a hazardous situation

Leveson (2009) believes that STPA *can be used at any stage of the system life cycle*. She summarizes STPA process in two main steps (Leveson 2009):

1. *Identifying potentially hazardous control actions*

There are four possibilities for a control action to be hazardous:

- *A control action required for safety is not provided or is not followed.*
- *An unsafe control action is provided that leads to a hazard.*
- *A potentially safe control action is provided too late, too early or out of sequence.*
- *A safe control action is stopped too soon (for a continuous or non-discrete control action) or applied too long.*

2. *Determining how unsafe control actions could occur*

In this step, causal scenarios are firstly created. Afterwards, the degradation of controls over time is analyzed. To identify the causal scenarios, causal factors should be defined for each component of the control loop (controlled process, sensor, controller and actuator, as shown in Fig. 2.4). The latest step of STPA is to analyze how controls could degrade over time and to consider protection barriers for the degradations.

A methodology is proposed for modeling the feedback network of CTSC project risks based on STAMP/STPA and system dynamics qualitative modeling. The methodology will be introduced and discussed in the next chapter.

References

De Rosnay J (1975) Le macroscope. Le Seuill, Paris
Dulac N (2007) A framework of dynamic safety and risk management modeling in complex engineering systems. PhD thesis submitted to the department of aeronautical and astronautical engineering at MIT, Feb 2007
Durand D (2010) La systémique. Presses Universitaires de France (PUF), 11th edn, Jan 2010
Forrester JW (1968) Industrial dynamics. The MIT press, 5th printing, May 1968
Forrester JW (1991) System dynamics and the lessons of 35 years. Massachusetts Institute of Technology. http://sysdyn.clexchange.org/people/jay-forrester.html
Fuchs HU (2006) System dynamics modeling in science and engineering. In: Invited talk at the system dynamics conference at the University of Puerto Rico, Resource center for science and engineering, Mayaguez, 8–10 Dec 2006
Garbolino E, Chéry JP, Guarnieri F (2009) Dynamic systems modelling to improve risk analysis in the context of Seveso industries. Chem Eng Trans 17:373–378
Garbolino E, Chéry JP, Guarnieri F (2010) Modélisation dynamique des systèmes industriels à risques, Sciences du Risques et du Danger (SRD), Lavoisier
GCCSI (2011) The global status of CCS: 2011, Global CCS Institute 2011, Canberra, Australia. ISBN 978-0-9871863-0-0
Hamblin WK, Christiansen EH (2004) Earth's dynamic systems. Prentice Hall, Pearson Education, Chap. 2, Geological systems
IRGC (2010) The emergence of risks: contributing factors. Report of International Risk Governance Council, Geneva, 2010. ISBN 978-2-9700672-7-6
Leveson N (2004a) A new accident model for engineering safer systems. Saf Sci 42:237–270
Leveson N (2004b) Model-based analysis of socio-technical risk, Massachusetts Institute of Technology, Engineering Systems Divison. Working Paper Series. ESD-WP-2004-08. Dec 2004
Leveson N (2009) Engineering a safer world, system safety for the 21st century. Massachusetts Institute of Technology, July 2009 (a draft book). http://sunnyday.mit.edu/safer-world.pdf. Accessed 1 Dec 2011
Louisot JP (2004) La gestion des risques dans les organisations applicable aux entreprises, aux collectivités territoriales & aux établissements de santé
Lyneis JM, Coopera KG, Els ShA (2001) Strategic management of Complex projects: a case study using system dynamics. Syst Dyn Rev 17(3):237–260. https://doi.org/10.1002/sdr.213
Meadows D, Meadows D, Randers J, Behrens W (1972) The limits to growth, New York
Morin E (2005) Introduction à la pensée complexe. Editions du Seuile, avril 2005
Pereira SJ, Lee G, Howard J (2006) A system-theoretic hazard analysis methodology for a non-advocate safety assessment of the ballistic missile defense system, June 2006. http://sunnyday.mit.edu/papers/BMDS.pdf
Powers WT (1990) Control Theory, a model of organisms, Syst Dyn Rev 6:1–20, ISSN 0883–7066
Radzicki MJ, Taylor RA (1997) Introduction to system dynamics. US Department of Energy http://www.systemdynamics.org/DL-IntroSysDyn/inside.htm
Richardson GP (2011) Reflections on the foundations of system dynamics. Syst Dyn Rev 27 (3):219–243. https://doi.org/10.1002/sdr.462

Richardson GP, Andersen DF (2010) Stakeholder dynamics. Paper presented at the 28th international conference of the system dynamics society, Seoul, Korea, July 25–29, 2010. http://www.systemdynamics.org/conferences/2010/proceed/index.html. Accessed Mar 2012

Rodrigues AG (2001) Managing and modelling project risk dynamics, a system dynamics-based framework. Paper presented at the 4th European project management conference, PMI Europe 2001, London UK, 6–7 June 2001

Samadi J, Garbolino E (2011) A new dynamic risk analysis framework for CO_2 Capture, Transport and Storage chain. In: 29th international conference of system dynamics society, 24–28 July 2011, Washington DC

Sterman JD (2000) Business dynamics, systems thinking and modeling for a complex world. The McGraw-Hill, USA

Sterman JD, Sweeney LB (2002) Cloudy skies: assessing public understanding of global warming. Syst Dyn Rev 18(2):207–240. https://doi.org/10.1002/sdr.242

Stringfellow MV (2010) Accident analysis and hazard analysis for human and organization factors. PhD thesis submitted to Massachusetts Institute of Technology, department of aeronautical and astronautical engineering, Oct 2010

Vensim Reference Manual (2010) Copyright 2010, Ventana Systems, Inc. Revision data August 22, 2010. http://www.vensim.com/. Accessed Feb 2011

Chapter 3
Systemic Methodology for Risk Management of CTSC Projects

Abstract The systemic methodology which is proposed for risk management of CTSC projects is introduced in this chapter. At the beginning, an overview of the methodology is presented. The methodology is founded on the concepts of STAMP and system dynamics. The objective is to model and analyze safety control structure involved in a CTSC project. Safety control structure is the organizational structure of stakeholders who are responsible for maintaining safety constraints. The goal of safety control structure in this work is to prevent CTSC projects delay or failure. This goal is rephrased as definition and treatment of significant risks that could avoid maintaining safety constraints. Following the identification of risks associated to CTSC projects progress, eighteen risks related to the phases prior to engineering are extracted. The aim is to put emphasis on the risks involved in the first phases of project development. Risks with different natures are selected and modeled by the proposed methodology. Stakeholders of CTSC projects are considered as the controllers. Required control actions for each controller (and for each particular risk) are discussed. Subsequently, inadequate control actions that could lead to a hazardous state are reviewed. System dynamics models are presented to understand the feedback networks affecting the amplification of each risk. Then, application of the methodology for three case studies (Barendrecht, Lacq and Weyburn) are explained. The context of each case study and major challenges related to each project are presented. Safety control structures are developed for each example in order to analyze the factors involved in the success or failure of projects. Afterwards, the three projects are compared in terms of context and associated risks. At the end of the chapter, a generic safety control structure is proposed for CTSC projects, according to the lessons learned from case studies analysis. Emphasis is placed on the importance of information feedback loops and communication between stakeholders, which lead to improve their mental models and decisions.

A methodology is proposed in this chapter to understand and analyze how risks could lead to success or failure of a CTSC project (risks have been previously presented in Chap. 1). The objective is to model and study the safety control structure involved in a CTSC project. Three case studies are reviewed and

J. Samadi and E. Garbolino, *Future of CO₂ Capture, Transport and Storage Projects*, SpringerBriefs in Environmental Science, https://doi.org/10.1007/978-3-319-74850-4_3

discussed. The methodology is based on the concepts of system dynamics and the systemic approach developed by Nancy Leveson at MIT, introduced in Chap. 2.

In this work, "Safety" is defined as the *absence of losses due to an undesired event (usually an accident)* (Leveson 1995). "Losses" in this definition include *human losses, mission or goal losses, equipment or material losses and environmental losses* (Dulac 2007). In this approach, *Safety is viewed as a dynamic control problem* (Leveson 2004).

The focus of the current work is on the mission or goal losses. Other kinds of failures could affect mission losses. The mission studied in this work is the success of a CTSC project.

CTSC is an emerging technology. Therefore, there is not a great amount of publicly available information on CTSC (CCP 2007), and even less on its organizational structure. In addition, most of available information on CTSC projects success or failure are extremely sensitive. Due to the confidentiality issues, there are not many publications on this subject.

However, the application of the methodology on this subject allows learning more about the complexity of CTSC projects risks. The required data are gathered from the available literature and project documents as well as discussions with the experts.

3.1 Overview of the Proposed Methodology

To study the safety control structure of CTSC projects, the following steps have been carried out. The purpose is to analyze the factors which make a CTSC project successful and the risks that prevent the project development.

1. Identifying major risks associated to CTSC (Table 3.1)
   (according to literature review, projects documentation and discussion with experts)
2. Assigning the risks to different CTSC subsystems and project phases (Figs. 3.1, 3.2 and 3.3)
3. Defining the nature of risks and their consequences (Fig. 3.4)
4. Extracting the risks related to the very first phases of the project (Table 3.2)
5. Modeling of CTSC projects safety control structure (Fig. 3.5)
6. Modeling the major risks using a systemic approach (Sect. 3.2)
7. Applying the systemic approach to model the safety control structure of different case studies (Sect. 3.3).

3.1.1 Step 1: Identifying Major Risks Associated to CTSC

A list of thirty-nine major risks are identified after reviewing several references (GCCSI 2009, 2011, 2016; Longannet 2011; Feenstra et al. 2010; Kerlero de Rosbo

Table 3.1 Overview of risks affecting CTSC project progress

Overview of risks affecting CTSC project progress			
1	Project permits not obtained	21	BLEVE
2	Technology scale-up	22	Lack of financial resources
3	Public opposition	23	Lack of political support
4	Lack of knowledge/qualified resources for operating the unit	24	Phase change and material problems
5	Corrosion	25	High cost of project
6	Using the existing facilities (specially pipelines)	26	Lower capture efficiency due to the upstream plant flexible operation
7	CO_2 out of specification	27	CO_2 leakage from compression unit
8	CO_2 plumes exceed the safe zone	28	Pipeline construction
9	Legal uncertainties	29	CO_2 leakage from pipeline
10	Safety related accident	30	Unavailability of regulations regarding different types of storage (offshore/onshore)
11	Uncertainties in stakeholders' requirements/perceptions—communication problems	31	Leakage through manmade pathways such as abandoned wells
12	Public availability of sensitive information	32	Well integrity
13	Change in policies/priorities	33	CO_2 migration
14	Financial crisis impact on financial support of CCS projects	34	Injectivity reduction over time
15	Unavailability of a monetary mechanism for CO_2	35	Uncertainties regarding the storage performance (capacity/injectivity/containment)
16	Construction field conditions	36	CO_2 leakage from storage to the surface
17	Geographical infrastructure	37	Model and data issues
18	Proximity to other industrial plants	38	Uncertainties related to storage monitoring
19	Energy consumption	39	Soil contamination
20	Maintenance and control procedures (including ESD system)		

2009; CCP 2007; Lacq Project 2012; Kapetaki and Scowcroft 2017). The list is available in Table 3.1.

The risks presented in this table are defined based on the project management definition of risk (*uncertain event or condition that, if it occurs, has a positive or negative effect on a project's objectives*) (PMBOK 2008).

3.1.2 Step 2: Assigning the Risks to Different CTSC Subsystems and Project Phases

The subsystem and project phase related to each risk are then identified (Figs. 3.1, 3.2 and 3.3). Risks are listed in the first column of the figures. The second column shows the related subsystem of each risk. "C", "T", "S" and "W" refer to "Capture", "Transport", "Storage" and the "Whole CTSC chain" respectively. In the other columns of the figures, affected project phases from each risk are defined.

3.1.3 Step 3: Defining the Nature of Risks and Their Consequences

At the third step, the nature of risks and their respective consequences are defined (Fig. 3.4). The nature of risk and risk consequences belongs to eight categories of risk already reviewed in Chap. 1. If we take "Project permits not obtained" as an example, the risk has a legal nature and therefore, risk nature is presented by "L" (Legal) in Fig. 3.4. Encountering such a risk will have consequences on the project and on global and local policies and strategies regarding CTSC. Consequently, "P" (Project) and "P/S" (Policy/Strategy) are specified as nature of consequences of "Project permits not obtained".

The second risk is "Technology scale up" which is a technical risk. Therefore, risk nature is presented by "T". Experiencing technology scale-up problems will affect the project progress. In addition, it may result not only in modifications of policies and strategies concerning CTSC technologies, but also in uncertainties about technical potential of CTSC to mitigate climate change. Hence, "P" (Project), "P/S" (Policy/Strategy) and "T" (Technical) are defined as nature of consequences for "Technology scale-up".

Figure 3.4 has to be read in this way.

3.1.4 Step 4: Extracting the Risks Related to the Very First Phases of the Project

At this stage, major risks associated to the very first phases of the project (before engineering) are extracted. The objective is to study the causes that prevent the project progress. The major risks are presented in Table 3.2.

	Risk	Subsystem[1]	Affected Project Phase					
			Opportunity	Definition and planning	Engineering	Construction	Operation (Injection of CO$_2$)	Post-injection (Monitoring)
1	Project permits not obtained	W	x	x	x	x	x	x
2	Technology scale-up	W	x	x	x	x	x	x
3	Public Opposition	W	x	x	x	x	x	x
4	Lack of knowledge/qualified resources for operating the unit	W	x	x	x	x	x	x
5	Corrosion	W			x	x	x	x
6	Using the existing facilities (specially pipelines)	W			x	x	x	x
7	CO$_2$ out of specification	W					x	x
8	CO$_2$ plumes exceed the safe zone	W					x	x
9	Legal uncertainties	W	x	x	x	x	x	x
10	Safety related accident	W				x	x	x
11	Uncertainties in stakeholders' requirements/perceptions - Communication problems	W	x	x	x	x	x	x
12	Public availability of sensitive information	W	x	x	x	x	x	x
13	Change in policies/priorities	W	x	x	x	x	x	x
14	Financial crisis impact on financial support of CCS projects	W	x	x	x	x	x	x
15	Unavailability of a monetary mechanism for CO$_2$	W	x	x	x	x	x	x

Fig. 3.1 Risks associated to CTSC and affected project phases. W = Whole CTSC chain, C = Capture, T = Transport, S = Storage

	Risk	Subsystem[1]	Opportunity	Definition and planning	Engineering	Construction	Operation (Injection of CO₂)	Post-injection (Monitoring)
				Affected Project Phase				
16	Construction field conditions	W				x		
17	Geographical infrastructure	W	x	x	x	x	x	x
18	Proximity to other industrial plants	W				x	x	x
19	Energy consumption	W					x	
20	Maintenance and control procedures (including ESD system)	W				x	x	x
21	BLEVE	W					x	
22	Lack of financial resources	W	x	x	x	x	x	x
23	Lack of political support	W	x	x	x	x	x	x
24	Phase change & material problems	W			x	x	x	x
25	High cost of project[2]	W	x	x	x	x	x	x
26	Lower Capture efficiency due to the upstream plant flexible operation	C					x	x
27	CO₂ leakage from compression unit	C					x	
28	Pipeline construction	T				x	x	x
29	CO₂ leakage from pipeline	T					x	

Fig. 3.2 Risks associated to CTSC and affected project phases. (1) W = Whole CTSC chain, C = Capture, T = Transport, S = Storage. (2) High cost is mostly due to capture and compression high costs (continued)

Risk	Subsystem[1]	Opportunity	Affected Project Phase				
			Definition and planning	Engineering	Construction	Operation (Injection of CO_2)	Post-injection (Monitoring)
30 Unavailability of regulations regarding different types of storage (offshore/onshore)	S	x	x	x	x	x	x
31 Leakage through manmade pathways such as abandoned wells	S					x	x
32 Well integrity	S					x	x
33 CO_2 migration	S					x	x
34 Injectivity reduction over time	S					x	
35 Uncertainties regarding the storage performance (capacity/injectivity/containment)	S	x	x	x	x	x	x
36 CO_2 leakage from storage to the surface	S					x	x
37 Model and data issues	S	x	x	x	x	x	x
38 Uncertainties related to storage monitoring	S	x	x	x	x	x	x
39 Soil contamination	S					x	x

Fig. 3.3 Risks associated to CTSC and affected project phases. W = Whole CTSC chain, C = Capture, T = Transport, S = Storage (continued)

	Risk	Subsystem[1]	Risk nature[2]	Nature of consequences[2]
1	Project permits not obtained	W	L	P, P/S
2	Technology scale-up	W	T	P, P/S, T
3	Public Opposition	W	S	P, P/S, L
4	Lack of knowledge/qualified resources for operating the unit	W	T, O/H	P, P/S, HSE, O/H, T
5	Corrosion	W	T	T, P
6	Using the existing facilities (specially pipelines)	W	T	T, P
7	CO_2 out of specification	W	T	T, P, HSE
8	CO_2 plumes exceed the safe zone	W	T	P, T, HSE
9	Legal uncertainties	W	L	P, P/S, T, L
10	Safety related accident	W	T, O/H	T, O/H, P, HSE, S
11	Uncertainties in stakeholders' requirements/perceptions- Communication problems	W	P, P/S, HSE, O/H, T, S, L, F/E	P, P/S, HSE, O/H, T, S, L, F/E
12	Public availability of sensitive information	W	O/H, P/S	P, P/S, S, O/H
13	Change in policies/priorities	W	P/S, L	P, P/S, L
14	Financial crisis impact on financial support of CCS projects	W	F/E	P, P/S, F/E
15	Unavailability of a monetary mechanism for CO_2	W	F/E, L	P, P/S, F/E, L
16	Construction field conditions	W	T	P, T
17	Geographical infrastructure	W	T	T, P, P/S, S, HSE
18	Proximity to other industrial plants	W	T	T, P, HSE
19	Energy consumption	W	T	P, P/S, T
20	Maintenance and control procedures (including ESD system)	W	T, O/H	T, O/H, P, HSE
21	BLEVE	W	T	P, T, HSE
22	Lack of financial resources	W	F/E	P, P/S, F/E
23	Lack of political support	W	P/S	P, P/S, O/H, S, L, F/E
24	Phase change & material problems	W	T	P, T
25	High cost of project[3]	W	F/E	P, P/S, F/E
26	Lower Capture efficiency due to the upstream plant flexible operation	C	T	P, T
27	CO_2 leakage from compression unit	C	T	T, P, HSE
28	Pipeline construction	T	T	P, T
29	CO_2 leakage from pipeline	T	T	T, P, HSE
30	Unavailability of regulations regarding different types of storage (offshore/onshore)	S	L	P, P/S, L
31	Leakage through manmade pathways such as abandoned wells	S	T	P, T
32	Well integrity	S	T	P, T
33	CO_2 migration	S	T	T, P, L, S
34	Injectivity reduction over time	S	T	P, T
35	Uncertainties regarding the storage performance (capacity/injectivity/containment)	S	T	P, P/S, T
36	CO_2 leakage from storage to the surface	S	T	T, P, P/S, HSE
37	Model and data issues	S	T	P, P/S, T
38	Uncertainties related to storage monitoring	S	T	P, P/S, T, HSE, S, L
39	Soil contamination	S	T	P, T

Fig. 3.4 Nature of CTSC risks and their consequences. (1) W = Whole CTSC chain, C = Capture, T = Transport, S = Storage. (2) T = Technical, P = Project, S = Social, P/S = Policy/Strategy, HSE = Health, Safety, Environment, L = Legal, O/H = Organizational/ Human, F/E = Financial/Economic. (3) High cost is mostly due to capture and compression high costs

Table 3.2 Major risks affecting the very first phases of the project

Major risks affecting CTSC project progress (in the first phases)			
1	Project permits not obtained	10	Unavailability of a monetary mechanism for CO_2
2	Technology scale-up	11	Geographical infrastructure
3	Public opposition	12	Lack of financial resources
4	Lack of knowledge/qualified resources for operating the unit	13	Lack of political support
5	Legal uncertainties	14	High cost of project
6	Uncertainties in stakeholders' requirements/perceptions— communication problems	15	Unavailability of regulations regarding different types of storage (offshore/onshore)
7	Public availability of sensitive information	16	Uncertainties regarding the storage performance (capacity/injectivity/containment)
8	Change in policies/priorities	17	Model and data issues
9	Financial crisis impact on financial support of CCS projects	18	Uncertainties related to storage monitoring

3.1.5   Step 5: Modeling of CTSC Projects Safety Control Structure

The modeling approach is developed based on the concepts of STAMP and system dynamics, introduced in Chap. 2. Modeling of CTSC projects safety structure is carried out within the framework of the following methodology which is composed of eight steps. The steps are schematized in Fig. 3.5.

1. The first stage is to define the goal of safety structure.

A major question about CTSC at the current stage of development is why some CTSC projects are successful to progress in particular contexts, while others fail? What are the main factors that affect the project progress?

   Therefore, the goal of safety structure defined in this work is to prevent the delay or cancelation of CTSC project.

   This objective could be interpreted as definition and treatment of significant risks that could prevent maintaining safety constraints.

   As Leveson (1995, 2004) affirms, there are four general ways to manage risks associated with a hazard:

- Eliminate the hazard from the system
- Reduce the hazard likelihood
- Assuring control measures when an undesired event is occurred
- Minimize damage in case of control measures absence

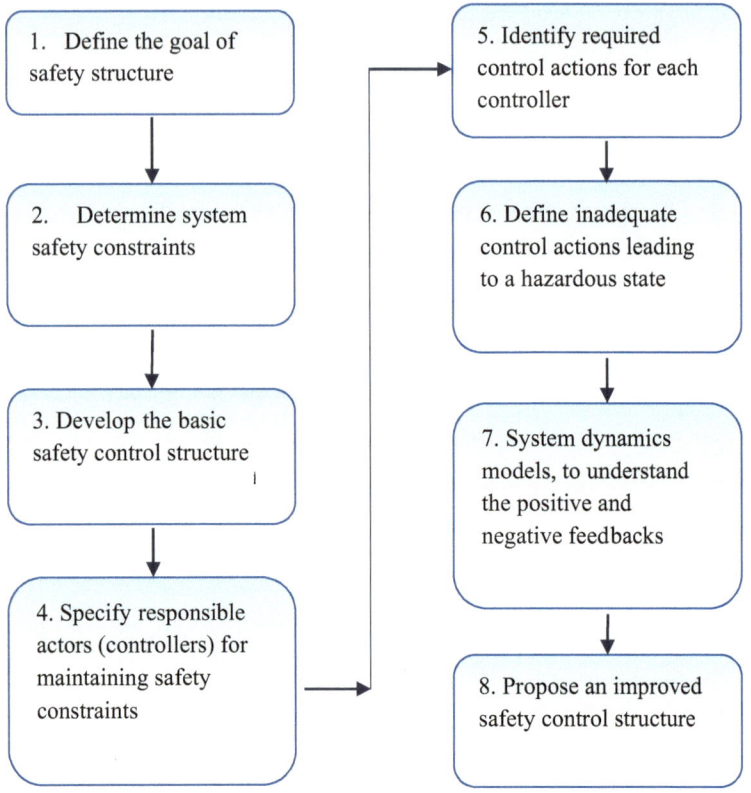

Fig. 3.5 Methodology of modeling CTSC projects safety control structure

2. In the second step, system safety constraints should be determined.

With the goal defined in the first step, the following constraints could be fixed for the system:

1st system safety constraint: The project must not be delayed or cancelled.

2nd system safety constraint: Measures of control must be provided in case of delay or cancellation.

In the next part, safety constraints will be detailed and analyzed for some major risks (defined in Table 3.2).

3. The basic safety control structure is developed in the third stage.

A general safety control structure has been previously presented in Chap. 2, Fig. 2.3.

The structure for CTSC is context specific, depending on several factors including location, population density and historic issues (CCP 2012). However, the following stakeholders are present in almost all cases:

- Project owner
- Politicians and Policy makers (National and Local)
- Regulators
- External experts
- Local population
- NGOs
- Media

Each of these stakeholders is a "controller" of the system, who is responsible for maintaining specified safety constraints.

4. A question needs to be answered at this level.

The question is who is responsible for maintaining each safety constraint?

For the safety constraints introduced in the second step, project owner is directly responsible. In other words, project owner is the endogenous controller, while other actors are exogenous controllers, who could affect the system and decisions of the project owner.

5. At this stage, required control actions for each controller should be identified.

Required control actions are the tasks that should be performed in order to maintain the safety constraints. These actions are risk specific.

6. Inadequate control actions that could lead to a hazardous state are defined in this stage.

Hazardous state is a state *that violates the safety constraints* (Leveson 2004).

Leveson presents four general types of inadequate control:

- *A required control action is not provided.*
- *An incorrect or unsafe control action is provided.*
- *A potentially correct or adequate control action is provided too late (at the wrong time).*
- *A correct control action is stopped too soon.*

7. System dynamics models, and especially causal graphs, are developed in this step.

The purpose is to study the positive and negative feedback loops which are involved in the process of maintaining safety constraints.

8. At the final step, an improved safety control structure is proposed based on the analysis of inadequate control actions and causal graphs.

3.2   Modeling Major Risks Affecting CTSC Project Progress

In this part, the specific safety constraints related to the risks are reviewed, and a number of these risks are modeled using the approach presented previously. Risks with different natures are selected in order to provide a more comprehensive model of risks.

The safety structures will not be discussed here. They will be presented and analyzed later for some case studies.

3.2.1   First Example: Risk of not Obtaining Project Permits

If we follow the modeling methodology schematized in Fig. 3.5, we need to start with defining the Safety Constraint.

Safety constraint: Required permits shall be obtained for Capture, Transport and Storage activities.

For understanding CTSC permitting procedures, a summary of significant points is provided here based on the report of CO_2 Capture Project on CTSC regulatory issues (CCP 2010).

Permitting requirements are not similar in different regions. There are two *generic approaches* for regulating CO_2 storage:

– *Integrated exploration and storage licensing frameworks.* This is the case of the EU.
– *Legislative amendments associated with existing oil and gas exploration legislation.* This is the case in Australia, Canada and a part of the US.

In the EU, the CCS Directive provides the legal framework for permitting CCS activities in the Member States. However, each country is interpreting the Directive to provide a national framework.

The US and Canada are finalizing their CO_2 storage legal frameworks. In the US, regulations are provided at the Federal level. In Canada, Federal and Provincial regulations for oil and gas are the basis of CTSC regulatory framework.

The EU CCS Directive determines two major permitting frameworks for CO_2 storage:

– The first one involves with the exploration phase, *where further information is needed to determine the suitability of the proposed site for CO_2 injection.* This stage takes between 6 and 24 months to be realized.
– The second one is associated with the storage permit. *A storage permit is a written decision by a Member State Competent Authority (CA) authorizing the geological storage of CO_2 in a suitable storage site by the operator. Permitting is not required for projects that are undertaken for research, development or*

testing of new products and processes. The storage threshold for the determination of such projects is 100,000 tonnes of CO_2 or less per year. Six to eight months are predicted for obtaining storage permit in the EU.

A planning process of 2–11 years is also expected. In this stage, Environmental Impact Assessment (EIA) is carried out.

The public and other third parties can influence the procedure by requesting additional information and by challenging information that has been presented. Therefore, in cases where there is public or third party opposition to the project, this stage of permitting process is particularly vulnerable to the risk of delay.

To understand the permitting procedure for Capture and Transport, the concept of "Carbon Capture Readiness" (CCR) should be reviewed. From 2009, all new combustion plants applying for operating permit in the EU have to be "CCS Ready". "CCS Ready" has been defined by different organizations such as IEA Greenhouse Gas R&D Program and GCCSI. Several points are still ambiguous in these definitions. However, the aim is to prove that CTSC technology could be introduced to the plant in the future.

Among the general stakeholders of CTSC project, project owner is responsible for maintaining the safety constraint related to the risk of not obtaining project permits.

If we continue the process by defining the required control actions and inadequate control actions leading to a hazardous state, the result could be as summarized in Table 3.3.

Table 3.3 Summary of first example, risk of not obtaining the required permits

Risk: Not obtaining the required permits
Safety Constraint:
Required permits shall be obtained for Capture, Transport and Storage activities.
Who is responsible for maintaining the safety constraint?
Project owner
Required Control Actions:
- Providing CCS Ready requirements - Requesting exploration permit if necessary - Carrying out the Environmental Impact Assessment (EIA) to obtain storage permit - Communicating with the public and other stakeholders in order to avoid oppositions which may lead to project delays
(Examples of) Inadequate Control Actions leading to a hazardous state:
- Environmental Impact Assessment is not provided. - Environmental authorities or public are not informed and consulted throughout the process. - Potential required changes of EIA report as the result of consultations are implemented too late. - Communication with the stakeholders is stopped too soon. (Therefore, all stakeholders' feedbacks could not be taken into consideration in the EIA report).

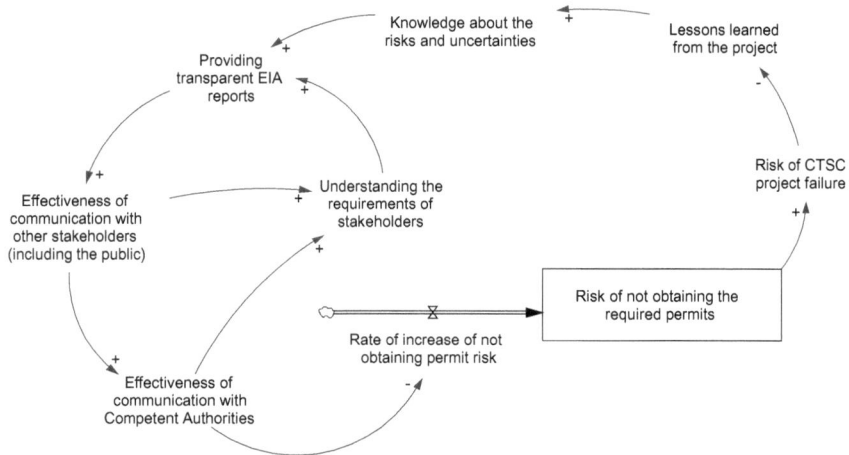

Fig. 3.6 Feedback network affecting the risk of not obtaining the required permits

Positive and negative feedbacks having an impact on the risk of not obtaining the required permits are shown in Fig. 3.6.

Risk of not obtaining the required permits is considered as a stock variable, since it is an accumulation in the system which we need to control. Rate of increase of the risk is a flow variable. Various control or auxiliary variables could lead to the modification of our flow variable. Effectiveness of communication with Competent Authorities reduces the risk of not obtaining the permits (negative feedback). Such effectiveness is a result of having effective communication with other stakeholders (including the public). A positive feedback loop is generated when the feedbacks from communication with stakeholders provide us with their requirements. As a result, more transparent EIA reports will be prepared, and the communication effectiveness will be increased consequently. Providing transparent EIA reports also requires knowledge on the risks and uncertainties. The knowledge could be improved by getting and analyzing lessons learned from the project. More lessons learned could be obtained if projects do not fail.

3.2.2 Second Example: Risk of Public Opposition

Poumadère et al. mention several points that drive CTSC public acceptance. Public perception of climate change, trust in industry and organizations in charge of project development, public participation from the very first phases of the project, history of the storage site, and socio-demographic characteristics of the local population (such as age, sex and level of higher education) are the major issues that stimulate the public to accept CTSC as a mitigation technology to deal with climate change (Poumadère et al. 2011).

Table 3.4 Summary of second example, risk of public opposition

Risk: Public opposition
Safety Constraints:
- Local population agreement should be assured.   - In case of opposition, measures should be in place to reduce the risk of project delay or cancellation.
Who is responsible for maintaining the safety constraint?
Project owner
Required Control Actions:
- Direct communication with the community from the initial phases of the project   - Giving information to the public in a less complicated manner (not too technical)   - Making the public trust them by highlighting the mutual benefits from the project development (including CTSC role in Climate Change mitigation)   - Making the public trust them by sharing the uncertainties and risks
(Examples of) Inadequate Control Actions leading to a hazardous state:
- Direct communication with the stakeholders is not provided.   - Communication with the stakeholders is performed indirectly (via media or third parties, for example).   - Direct communication with the stakeholders is provided too late.   - Project developers do not continue to directly communicate with the stakeholders during the life cycle of the project.

The safety constraints for public opposition risk are as follows:

Safety constraint 1: Local population agreement should be assured.

Safety constraint 2: In case of opposition, measures should be in place to reduce the risk of project delay or cancellation.

Project owner is responsible to ensure and provide the required supports for maintaining safety constraints.

The next steps of defining the required control actions and inadequate control actions leading to a hazardous state, lead us to the summary shown in Table 3.4.

Figure 3.7 summarizes the variables involved in the control process of public opposition risk.

In order to reduce the rate of public opposition risk, more effective communication has to be ascertained. Once more, improving our knowledge through the lessons learned will increase our willingness to share the information with the stakeholders and among them the local community. Sharing the information will make the public trust the project owner and other stakeholders. In addition, public perception of climate change will be improved. As previously mentioned, the history of the storage site is a significant factor for assuring public acceptance.

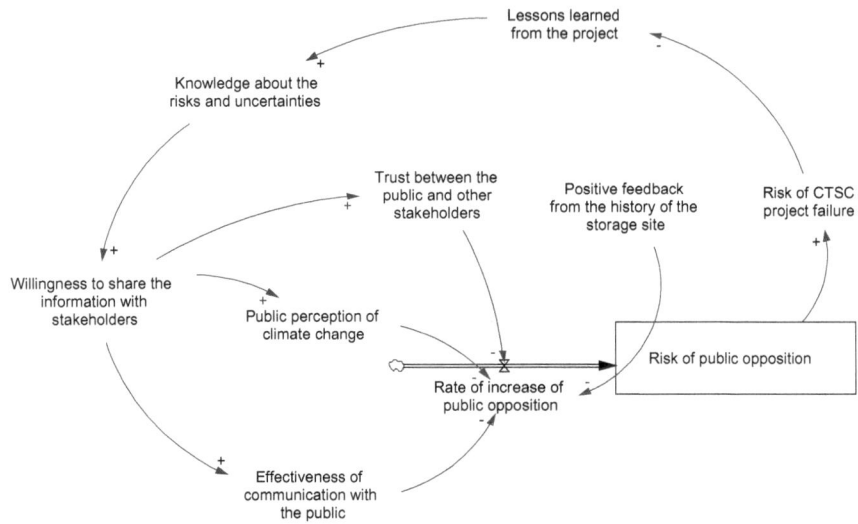

Fig. 3.7 Feedback network affecting the risk of public opposition

3.2.3 Third Example: Risk of Financial Resource Shortage

Financial support is essential to have commercial scale CTSC projects. Several projects have been stopped due to financial resource problems. Longannet project in the United Kingdom is an example. The project was cancelled in October 2011, since it was not affordable, and stakeholders' risk perceptions were different (Thomas et al. 2012; GCCSI 2012). On June 26, 2012, Peel Energy project in the UK was cancelled due to *the economic slowdown and uncertainties around public funding* (GCCSI 2012). Recent data bases confirm that financial problems are a major reason of CTSC projects cancellation (MIT 2016).

The safety constraint for the risk of financial support shortage could be formulated as follows:

Safety constraint: Financial support shall be ensured for commercial scale CTSC projects.

Government and project owner are responsible to maintain the safety constraint.

Defining the required control actions and inadequate control actions leading to a hazardous state, lead us to the summary shown in Table 3.5.

The feedback network affecting lack of financial resource is showed in Fig. 3.8.

Rate of financial support risk is directly affected by local policy of each region about CTSC. Local policy and national/international policies are mutually interconnected. Lessons learned and knowledge about the risks and uncertainties will have an effect upon policies. The policies about CTSC determine whether funds will be allocated for the CTSC project. Furthermore, correct estimation of required

Table 3.5 Summary of third example, risk of financial resource shortage

Risk: Financial resource shortage
Safety Constraint:
Financial support shall be ensured for commercial scale CTSC projects.
Who is responsible for maintaining the safety constraint?
Government & Project owner
Required Control Actions:
For government: - Providing the necessary financial support for the project And for project owner: - Correctly estimating the required financial resource in the feasibility studies - Allocating the received financial support for the project development
(Examples of) Inadequate Control Actions leading to a hazardous state:
- Cost estimation is not performed correctly at the initial phases. - Effect of some parameters is not taken into account in the first cost estimations. - Realistic cost estimations are provided too late. - Feedbacks from external experts/suppliers are not completely integrated in project cost calculations.

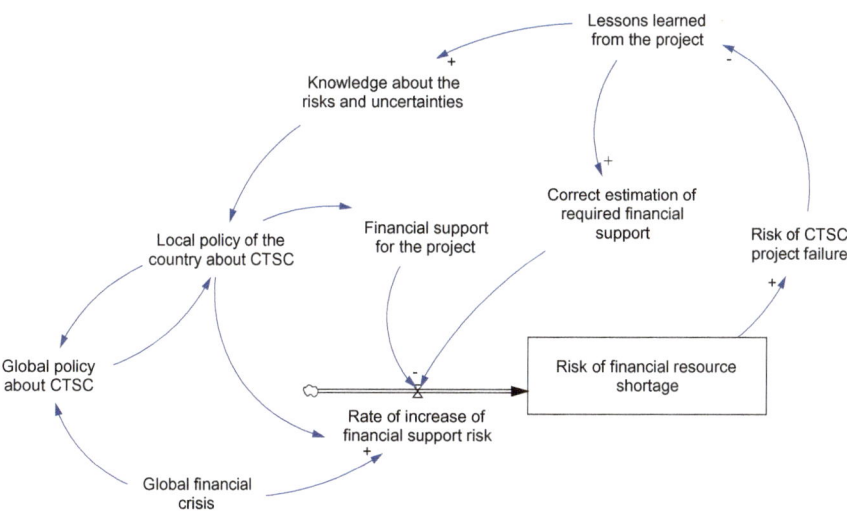

Fig. 3.8 Feedback network affecting the risk of financial resource shortage

financial support could be assured by the lessons learned from the project. As illustrated in Fig. 3.8, the feedbacks of knowledge/policy, local/global policies, policy/financial support, policy/risk of financial lack and cost estimation/risk of financial lack do not have any positive or negative sign. This is due to the uncertainties about whether knowledge improvement on CTSC could lead to change the policies to more or less investment on CTSC technology. These uncertainties are formulated by Tombari as "learning curve" uncertainty. The idea is that we are not sure if learning from CTSC projects results in getting less expensive technologies. The notion of learning curve comes from Schlumberger Carbon Services, who believes that First Of A Kind (FOAK) CTSC plants will experience a *"pre-learning" phase, in which cost decreases will not be uniform.* It is argued that immature technologies often go through this phase which is *commonly referred to as the "valley of death". In order to advance, the technology requires more and more funding with riskier returns* (Tombari 2011; Soupa et al. 2012). In addition, global financial crisis has an influence on global policy about CTSC. Positiveness or negativeness of the feedback is uncertain at the moment.

3.2.4 Risk Interconnections

As discussed earlier, CTSC is a novel complex technology for which risks cannot be analyzed and managed separately. The interrelations of risks create a context which has the potential to give rise to a hazardous state. Therefore, the interconnections of risks shall be modeled and studied. Major risks affecting CTSC project development were introduced in Table 3.2. Inter-relations of these risks are illustrated in Fig. 3.9. The green bold feedbacks represent the risks interconnections.

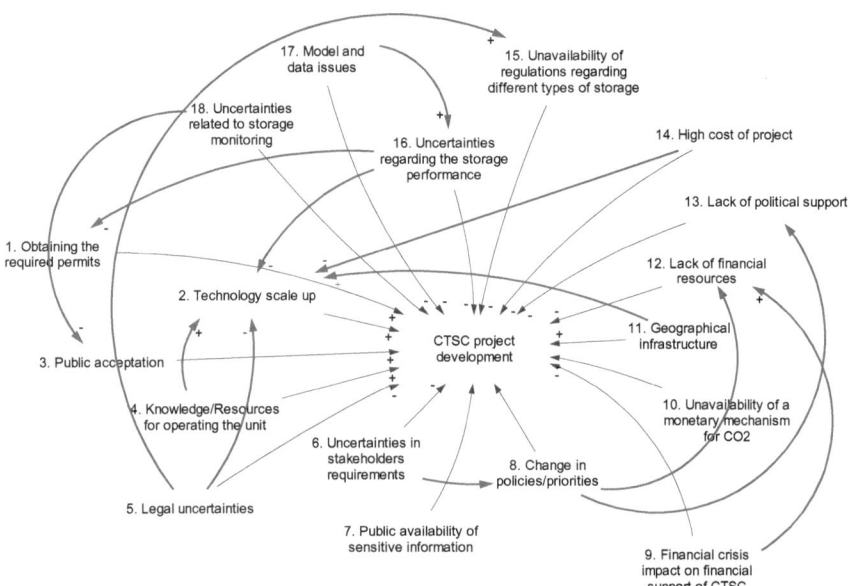

Fig. 3.9 Interconnections of major risks affecting CTSC projects progress

An example of risk interconnections is "Technology scale up" which is influenced by five other risks: "Knowledge/Resources for operating the unit", "Legal uncertainties", "Geographical infrastructure", "High cost of project" and "Uncertainties regarding the storage performance".

3.3   Application of the Methodology for Case Studies

In this section, application of the methodology for three case studies is explained and discussed. The case studies are selected based on the level of project success.

The aim is to analyze the context and safety control structure of different projects to find the rules and elements leading to the progress of CTSC projects to commercial scales.

The first example is Barendrecht, in the Netherlands, which was cancelled due to public opposition and lack of local support.

The second example is Lacq, as the first CTSC pilot plant in France, in which CO_2 injection has been done in spite of some technical challenges.

The third example is Weyburn, as a successful industrial scale EOR project in the North America, which has to deal with some questions.

As noted before, going through details of case studies is impossible because of lack of information.

3.3.1   First Example: Barendrecht

Barendrecht was a CTSC integrated project, planned to inject 400,000 tonnes CO_2 per year. CO_2 was produced in a hydrogen production plant and planned to be injected in two depleted gas fields. The capture plant is located about 20 km from Barendrecht, a town located in the west of the Netherlands. Barendrecht is situated at around 14 km of Rotterdam, the second largest city in the Netherlands. The population of the city is about 44,000 people.

A pipeline of 16.5 km was designed to transport the captured CO_2 to the storage location. The first gas field (Barendrecht) could store about 0.8 million tonnes of CO_2 at a depth of 1700 m. The second gas field (Barendrecht-Ziedewij) could store about 9.5 million tonnes of CO_2 at a depth of 2700 m.

Shell was the owner of the project, and a financial support of 30 million euros was invested by the government for this project. Shell would also have the benefit of emission saving under ETS (Emissions Trading System) program.

The tender was announced by the Dutch government in 2007. In early 2008, Shell was selected as the winner of the tender. Debates have begun from then on, when the project was presented to local community. The first phase of injection was planned to start in 2011 for a duration of three years. Injection in the second gas field was planned to begin in 2015 for 25 years (Feenstra et al. 2010).

In November 4th, 2010, Dutch Minister of Economic Affairs, Agriculture and Innovation announced that the project is cancelled. *The delay of the CO_2 storage project for more than 3 years and the complete lack of local support are the main reasons to stop.* However, the minister believes that Barendrecht experiences are valuable for further development of CO_2 storage in the Netherlands. So, *Barendrecht cancellation does not mean the end of CO_2 storage in the Netherlands* (CCJ 2010; Netherlands Government 2010).

Here we will discuss the application of the methodology for Barendrecht project. The purpose is to understand the weaknesses of the project safety structure, and the points that could be improved to avoid the delay and stop.

The first two steps of the approach presented in Fig. 3.5 are the same as the ones discussed earlier in this chapter. Therefore, the central point of discussion in this part is the actors who play a role in the progress of the project.

In the following paragraphs, the main stakeholders and their responsibilities are summarized (Feenstra et al. 2010):

– **National government**: was engaged via two ministers: Ministry of Economic Affairs (EZ) and Ministry of Housing, Spatial Planning and Environment (VROM).
   EZ established a group (Task force CCS), with representatives of industry, NGOs and local governments, to support CTSC development in the Netherlands.
– **Local governments**: were involved at two levels: provincial and municipal.
   The executive board of the provincial government was responsible for environmental permitting procedures. An environmental protection agency (DCMR) was appointed by the provincial deputy to execute the leadership of a consultation group (BCO2). BCO2 was the administrative consultation group of Barandrecht project.
   At the municipal level, governments of Barendrecht and Albrandswaard were involved. Albrandswaard population did not raise many concerns about the project, probably because a few numbers of their houses were located directly above the gas fields. Barendrecht government was more actively involved.
– **Project developers**: Three companies were engaged. Shell was the initiator and responsible for storage and monitoring. Two other companies were collaborating with Shell for capture and transport. NAM (Netherlandse Aardolie Maatschappij BV), the Netherlands biggest oil and natural gas producer, was *responsible for existing natural gas production from the gas fields in Barendrecht.* OCAP (Organic CO_2 for Assimilation of Plants) was responsible for CO_2 transport.
– **External experts, consultants and research organizations**: Several external experts were involved, mainly by project developers, for environmental studies of CO_2 storage and to answer the questions from municipality in the public meetings.
– **NGOs**: Several NGOs were also active for or against the project. Greenpeace is opposed to CTSC, at national and international scales. Uncertainties about

subsurface capacity to store CO_2, energy waste, risk of CO_2 leakage and expensiveness are the principal concerns of Greenpeace regarding CTSC technology (Rochon et al. 2008). SNM, the Netherlands Society for Nature and Environment, believes that CTSC is essential as an intermediate step towards clean energy.

- **Local population**: The people who live in the neighborhood of CO_2 storage location are significant stakeholders of CTSC projects. In Barendrecht case, they were represented by the municipal government.
- **Media**: Local and national newspapers, as well as televisions, websites and magazines were another actors who were involved in distributing information on the project.

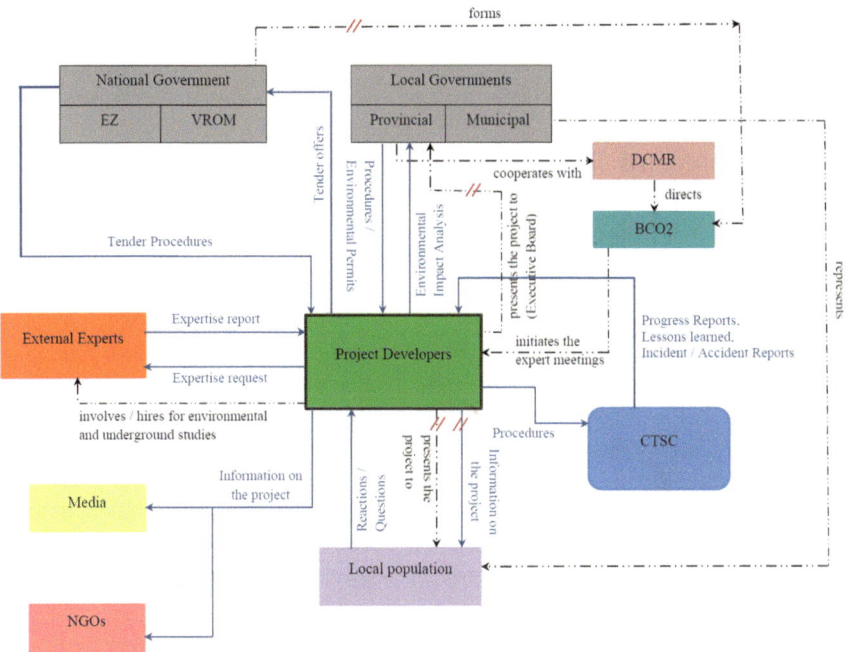

Fig. 3.10 Barendrecht safety control structure, initial model

LEGEND

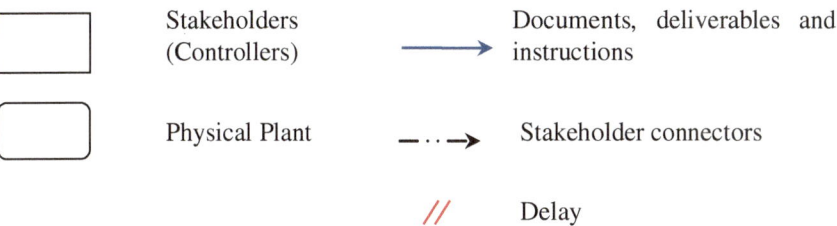

The organizational structure of Barendrecht project is illustrated in Fig. 3.10.

Rectangles with sharp corners symbolize the stakeholders (controllers), while the round-corner rectangle (CTSC) stands for the physical plant (same legend as introduced in Leveson 2004).

Dash lines are the stakeholders' connectors, which show the relations of actors. Documents, deliverables and instructions exchanged between the actors are represented by solid lines.

// on the arrows represents delay, which is also a system dynamics concept.

When delay symbol (//) is put on a connection, it means that the action is carried out with delay.

Lessons learned from the project confirm that communication problems are the main issues resulted in the opposition to the project. The most significant subjects affecting the effectiveness of the safety control structure are as following:

1. As showed in Fig. 3.10, there is no connection between the national and local governments. The lack of such connection reinforced the public opposition.
2. Delays in some required actions made the community resist to the project. Some examples are presented in Fig. 3.10. Establishment of the administrative consultation group (BCO2) by the national government occurred rather belatedly, after the start of local opposition. Delay symbol on the connection between National Government and BCO2 illustrates such late reaction. In addition, presentation of the project to the community (Local Governments and population) happened with delay. Some information on the project was not communicated upon request, especially due to confidentiality issues.
3. Regulatory responsibilities were not so clear in the project context. Changing the project regulatory framework was another reason for which the opposition occurred. In the new framework, the project would be considered as a one having national impacts. Therefore, National Government was authorized for all needed permissions, even those normally awarded by local governments.
4. Another issue is the lack of mutual connection between the stakeholders in some cases. For example, feedbacks of local governments were not taken into consideration by the project developers, although the project had been presented to the local community.

   In some cases, mutual connections are not available for a particular reason. For instance, NGOs preferred to announce their opinion in the national level, instead of on this specific project. Therefore, no feedback is considered from NGOs to the project developers. The media also tried not to influence opinions. Thus, no direct connection is available from the media to the project developers.

So, the project safety control structure could be improved as presented in Fig. 3.11. The added elements are presented in orange. Delays existed in Fig. 3.10 are removed in the proposed improved model (Fig. 3.11).

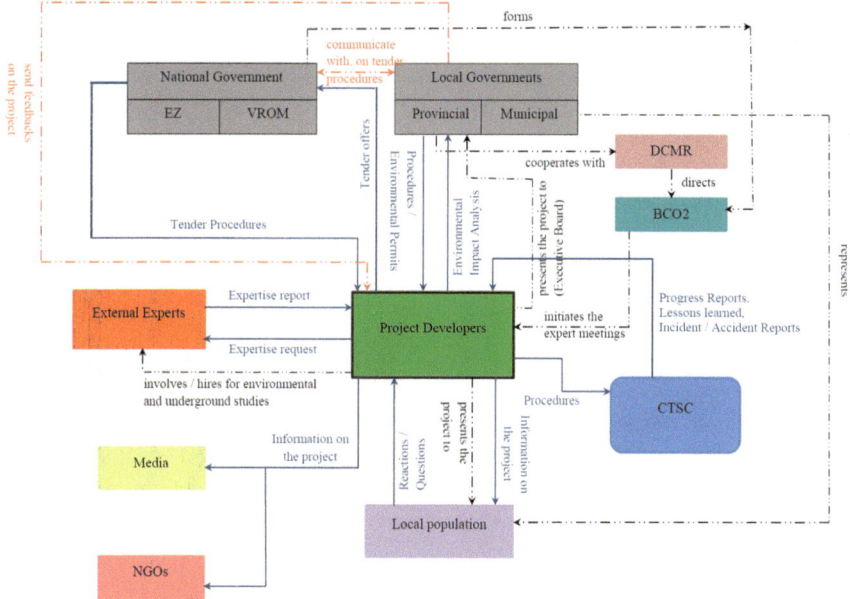

Fig. 3.11 Barendrecht safety control structure, improved model

LEGEND

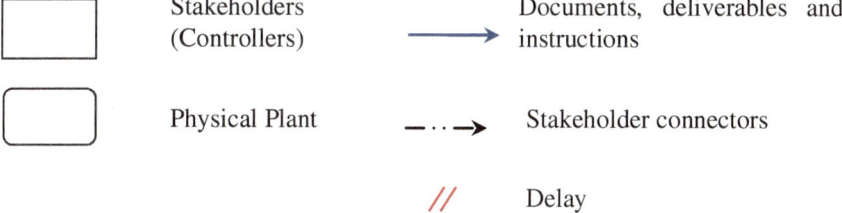

☐ Stakeholders (Controllers)	→	Documents, deliverables and instructions
☐ Physical Plant	–··→	Stakeholder connectors
	//	Delay

3.3.2 Second Example: Lacq

Lacq is a CTSC integrated pilot project in France to inject 120,000 tonnes CO_2 in a depleted gas reservoir (at a depth of 4500 m) during two years. The storage site is planned to be monitored during three years after the end of injection. Following the monitoring phase, the responsibility will be transferred to the government. It means that the project owner will not be responsible after these five years.

CO_2 is produced in a natural gas production unit which is situated in Lacq, a city in the South west of France in Pyrénées-Atlantiques region.

An existing pipeline of 29 km transports CO_2 to the injection location, which is located in 3 km of Jurançon city. Around 7000 people live in Jurançon [7087 in

2004 (Mairie Jurançon 2012)]. The Capture plant comes within ICPE regulation. The pipeline and the injection site are under the mining code.

In February 8, 2007, Total (the project owner) announced the decision of performing Lacq CTSC pilot plant in a news conference. From 6th to 30th of November 2007 a public dialogue was taken place to inform the local stakeholders on the project and understand their points of view and concerns (C&S Conseils 2008).

The regional government asked the project owner to conduct a public survey before giving the permits for the project start-up. A public survey was conducted for 64 days, from July 21, 2008 to September 22, 2008. A positive opinion on the project was given by the survey committee (at the end of October 2008) following the results of the survey.

On May 13, 2009, a decree was published by the regional prefecture to authorize the start of the project.

The injection was started in January 8, 2010 and planned to be terminated on April 2012. On September 12, 2011, Total requested an extension of 18 months for the injection, due to the technical problems of some equipment. In April 2011, 23,000 tonnes CO_2 was injected into the reservoir, while the objective was to inject 75,000 tonnes CO_2 (CLIS 2011).

The major stakeholders of the project are as follows:

- **Regional (Local) Government**: Several representatives of the regional government are involved, including the prefects and DRIRE (Direction Régionale de l'Industrie, de la Recherche et de l'Environnement). Mayors and deputy mayors of different communities are also engaged.
  DRIRE is a French governmental structure which is responsible for controlling the regulative compliance of the installation in ICPE framework (for ICPE definition, refer to Chap. 1) (ICPE website 2). Since January 2010, DRIRE has been merged with two other structures, DIREN (Direction Régionale de l'Environnement) and DRE (Direction Régionale de l'Equipement). These three merged structures form DREAL (Direction Régionale de l'Environnement, de l'Aménagement et du Logement). DREAL is conducted by the Ministry of Ecology, Energy and Sustainable Development (MEEDDM: Ministère de l'Ecologie, de l'Energie, du Développement Durable et de la Mer).
  A local committee (CLIS: Commission Locale d'Information et de Suivi) has been created by the regional prefecture to follow up the project progress. Regular meetings have been held since June 2008, when CLIS was established.
- **Project Owner**: Total is the owner of the project. Some other companies cooperate with Total, such as Air Liquide for the oxycombustion unit.
- **External experts**: from universities and research organizations have been requested to verify whether there are significant environmental and health risks concerning the project. If so, preventive and protective barriers for the potential risks were asked to be identified. The experts also seek to improve their knowledge on the possibility of commercial scale CTSC projects in France.
- **NGOs**: Several environmental NGOs have participated in the debates since the first public presentation of the project. An external specialist was asked by one

of the NGOs to evaluate the project. Having one single private firm (Total) as the owner of the project is a main issue raised by the expert. He believes that for such a project, which has a life cycle much more than the company's life cycle, organizations working on long term monitoring and risk management have to contribute (CLIS 2008).
- **Local population**: is again a main stakeholder of the project.
- **Media**: Local and national newspapers and websites spread the information concerning the project.

The organizational structure of Lacq project is illustrated in Fig. 3.12.

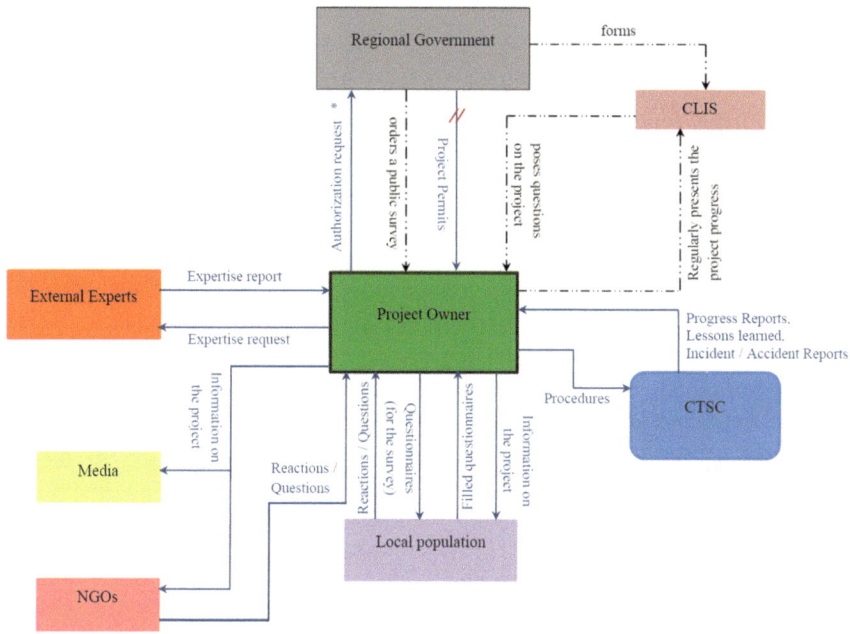

* Includes Environmental Impact Assessment, Hazard Analysis and HSE issues

Fig. 3.12 Lacq safety control structure, initial model

LEGEND

☐	Stakeholders (Controllers)	⟶	Documents, deliverables and instructions
☐	Physical Plant	– .. ⟶	Stakeholder connectors
		//	Delay

Delay of the regional government to give the permits is due to the required time for consulting different organizations and obtaining their opinion on the project. It could last between 10 and 12 months (ICPE website 3). Principal questions of CLIS from the project owner contain:

– The monitoring system of the project
– If the available protection barriers are sufficient to protect the local population
– The role of scientific committee (external experts) regarding the project.

Contrary to the Barendrecht case, there is a lack of published information on Lacq organizational structure. Therefore, an improved safety control structure cannot be proposed for this particular case study.

A general improved structure will be suggested at the end of the chapter, when the case studies are discussed.

3.3.3 Third Example: Weyburn

Weyburn is an oil field located in both Canada and the United States. The aim is to verify the feasibility of CO_2 geological storage under an Enhanced Oil Recovery (EOR) research project. The CO_2 is a byproduct of Dakota Gasification Company's synthetic fuel plant in North Dakota. The CO_2 is purchased from the fuel plant and is transported to Williston basin (Weyburn is a part of this basin) through a pipeline of 320 km. The first phase of injection was started on September 15, 2000. The initial injection rate was 5000 tonnes/day, and about 20 million tonnes of CO_2 is expected to be injected into the reservoir. Weyburn is a 180 km^2 oil field discovered in 1954. It is estimated that the oil production will increase by 130 million barrels (10% of the original oil in place) through the EOR operations. The oil field life is estimated to be increased by 25 years (PTRC 2004; Verdon 2012).

The project was launched by PTRC (Petroleum Technology Research Center), in Regina, Saskatchewan, in collaboration with Encana (now Cenovus) in Calgary, Alberta. The fund is provided by several governments and industries of Canada, the United States, Europe and Japan (PTRC 2004).

In January 2011, a farmer couple, having their land over the Weyburn CO_2 storage site, claimed that the injected CO_2 has been leaked, killed animals and *sent groundwater foaming to the surface like shaken-up soda-pop*. They asked a consultant (Petro-Find) for a soil gas study. The results showed that the source of CO_2 high concentrations in the soil is the injected CO_2 into the Weyburn reservoir (CBC news 2011).

PTRC and Cenovus, the project owners, called for an independent expertise. They announced that no leakage has been identified in the Weyburn field, and the source of CO_2 claimed by the farmers is not the Weyburn reservoir (Whittaker 2011). However, Ecojustice (a Canadian Environmental NGO) claims that there are important unanswered questions in PTRC response to the soil gas studies (Ecojustice 2011). In March 2011, Petro-Find performed another soil gas survey, and confirmed that the source of CO_2 found in the soil gas is the *anthropogenic CO_2 injected into the Weyburn reservoir* (Lafleur 2011).

In spite of debates on the leakage, the project is still in operation (GCCSI website 2017).

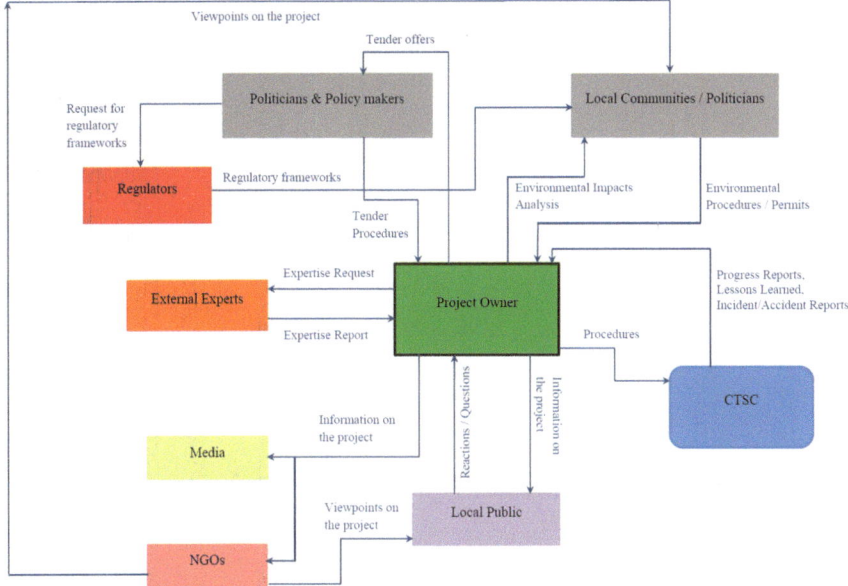

Fig. 3.13 Weyburn safety control structure, rough model based on CCP (2012)

LEGEND

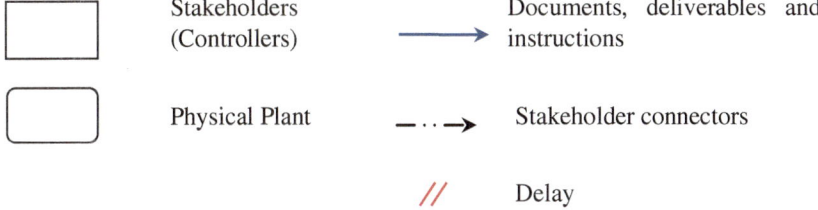

Details on Weyburn project stakeholders are not available. The following structure (Fig. 3.13) is prepared based on CCP (2012), which is an industry point of view of stakeholders.

Similar to the case of Lacq, a great amount of information, especially on the organizational issues, are confidential, and consequently unavailable on Weyburn project.

In order to propose an optimized safety control structure for CTSC projects, we will firstly analyze the case studies in subsequent sections.

3.4 Comparison of Case Studies, from Context Point of View

As discussed earlier in previous chapters, risks are emergent properties of systems and therefore, have to be analyzed by taking into account the context in which they are generated. In addition, CTSC projects safety control structure is context specific

Table 3.6 Comparison of case studies' context

	Barendrecht	Lacq	Weyburn
Current status	Cancelled (in detailed organization phase)	Monitoring	In operation
Scale	Demonstration	Pilot	LSIP
CO_2 storage rate	400,000 tonnes/year	60,000 tonnes/year	3 Mtpa
Storage type	Depleted gas field	Depleted gas field	EOR
Country	The Netherlands	France	The United States
Major issues	Public opposition	Technical challenges	– Public acceptance challenges – EOR as a long term storage option!
Main objective	Set down a foundation for CTSC LSIP in the Netherlands	Verify the feasibility of a CO_2 storage plant in France	Oil production increase
Concerning industry	Oil and Gas	Oil and Gas	Oil and Gas

and depends on several factors. For these reasons, it is essential to compare the case studies in terms of context.

The context comparison of Barendrecht, Lacq and Weyburn projects are presented in Table 3.6.

3.5   Comparison of Case Studies, from Risk Point of View

The results of comparing the three case studies in terms of CTSC project risks are presented in the following figures.

Figure 3.14 (risks 1–18) contains the risks concerning the phases prior to engineering. Figure 3.15 (risks 19–39) includes the remainder.

Barendrecht was cancelled in the first phases of its progress. Consequently, the second group of risks is irrelevant to Barendrecht. The (potential) risks involved in the context of Lacq and Weyburn are much more numerous since these projects are in advanced project phases.

Barendrecht	Lacq	Weyburn
☒ 1. Project permits not obtained	☐ 1. Project permits not obtained	☐ 1. Project permits not obtained
☒ 2. Technology scale-up	☒ 2. Technology scale-up	☐ 2. Technology scale-up
☒ 3. Public Opposition	☒ 3. Public Opposition	☒ 3. Public Opposition
☐ 4. Lack of knowledge/qualified resources for operating the unit	☒ 4. Lack of knowledge/qualified resources for operating the unit	☐ 4. Lack of knowledge/qualified resources for operating the unit
☒ 5. Legal uncertainties	☒ 5. Legal uncertainties	☒ 5. Legal uncertainties
☒ 6. Uncertainties in stakeholders requirements/perceptions - communication problems	☒ 6. Uncertainties in stakeholders requirements/perceptions - communication problems	☒ 6. Uncertainties in stakeholders requirements/perceptions - communication problems
☒ 7. Public availability of sensitive information	☒ 7. Public availability of sensitive information	☒ 7. Public availability of sensitive information
☒ 8. Change in policies/priorities	☒ 8. Change in policies/priorities	☐ 8. Change in policies/priorities
☐ 9. Financial crisis impact on financial support of CCS projects	☐ 9. Financial crisis impact on financial support of CCS projects	☐ 9. Financial crisis impact on financial support of CCS projects
☐ 10. Unavailability of a monetary mechanism for CO_2	☐ 10. Unavailability of a monetary mechanism for CO_2	☐ 10. Unavailability of a monetary mechanism for CO_2
☐ 11. Geographical infrastructure	☐ 11. Geographical infrastructure	☐ 11. Geographical infrastructure
☐ 12. Lack of financial resources	☐ 12. Lack of financial resources	☐ 12. Lack of financial resources
☒ 13. Lack of political support	☐ 13. Lack of political support	☐ 13. Lack of political support
☐ 14. High cost of project	☐ 14. High cost of project	☐ 14. High cost of project
☐ 15. Unavailability of regulations regarding different types of storage (offshore/onshore)	☐ 15. Unavailability of regulations regarding different types of storage (offshore/onshore)	☐ 15. Unavailability of regulations regarding different types of storage (offshore/onshore)
☐ 16. Uncertainties regarding the storage performance (capacity/injectivity/containment)	☒ 16. Uncertainties regarding the storage performance (capacity/injectivity/containment)	☐ 16. Uncertainties regarding the storage performance (capacity/injectivity/containment)
☐ 17. Model and data issues	☒ 17. Model and data issues	☒ 17. Model and data issues
☐ 18. Uncertainties related to storage monitoring	☒ 18. Uncertainties related to storage monitoring	☒ 18. Uncertainties related to storage monitoring

Fig. 3.14 Comparison of risks associated to case studies

Barendrecht	Lacq	Weyburn
□ 19. Corrosion	☒ 19. Corrosion	☒ 19. Corrosion
□ 20. Using the existing facilities (specially pipelines)	☒ 20. Using the existing facilities (specially pipelines)	□ 20. Using the existing facilities (specially pipelines)
□ 21. CO_2 out of specification	☒ 21. CO_2 out of specification	☒ 21. CO_2 out of specification
□ 22. CO_2 plumes exceed the safe zone	☒ 22. CO_2 plumes exceed the safe zone	☒ 22. CO_2 plumes exceed the safe zone
□ 23. Safety related accident	☒ 23. Safety related accident	☒ 23. Safety related accident
□ 24. Construction field conditions	□ 24. Construction field conditions	□ 24. Construction field conditions
□ 25. Proximity to other industrial plants	□ 25. Proximity to other industrial plants	□ 25. Proximity to other industrial plants
□ 26. Energy consumption	☒ 26. Energy consumption	☒ 26. Energy consumption
□ 27. Maintenance and control procedures (including ESD system)	☒ 27. Maintenance and control procedures (including ESD system)	☒ 27. Maintenance and control procedures (including ESD system)
□ 28. BLEVE	□ 28. BLEVE	□ 28. BLEVE
□ 29. Phase change & material problems	☒ 29. Phase change & material problems	☒ 29. Phase change & material problems
□ 30. Lower Capture efficiency due to the upstream plant flexible operation	☒ 30. Lower Capture efficiency due to the upstream plant flexible operation	☒ 30. Lower Capture efficiency due to the upstream plant flexible operation
□ 31. CO_2 leakage from compression unit	☒ 31. CO_2 leakage from compression unit	☒ 31. CO_2 leakage from compression unit
□ 32. Pipeline construction	□ 32. Pipeline construction	□ 32. Pipeline construction
□ 33. CO_2 leakage from pipeline	☒ 33. CO_2 leakage from pipeline	☒ 33. CO_2 leakage from pipeline
□ 34. Leakage through manmade pathways such as abandoned wells	☒ 34. Leakage through manmade pathways such as abandoned wells	☒ 34. Leakage through manmade pathways such as abandoned wells
□ 35. Well integrity	☒ 35. Well integrity	☒ 35. Well integrity
□ 36. CO_2 migration	☒ 36. CO_2 migration	☒ 36. CO_2 migration
□ 37. Injectivity reduction over time	☒ 37. Injectivity reduction over time	☒ 37. Injectivity reduction over time
□ 38. CO_2 leakage from storage to the surface	☒ 38. CO_2 leakage from storage to the surface	☒ 38. CO_2 leakage from storage to the surface
□ 39. Soil contamination	☒ 39. Soil contamination	☒ 39. Soil contamination

Fig. 3.15 Comparison of risks associated to case studies (continued)

Interactions of risks associated to each case study are presented in Figs. 3.16, 3.17, 3.18 and 3.19. The risks for which evidences/references are available are highlighted in green bold, while risks having the potential to affect the projects are represented in violet. To avoid models' complexity, interactions of all thirty-nine risks are not shown in the figures.

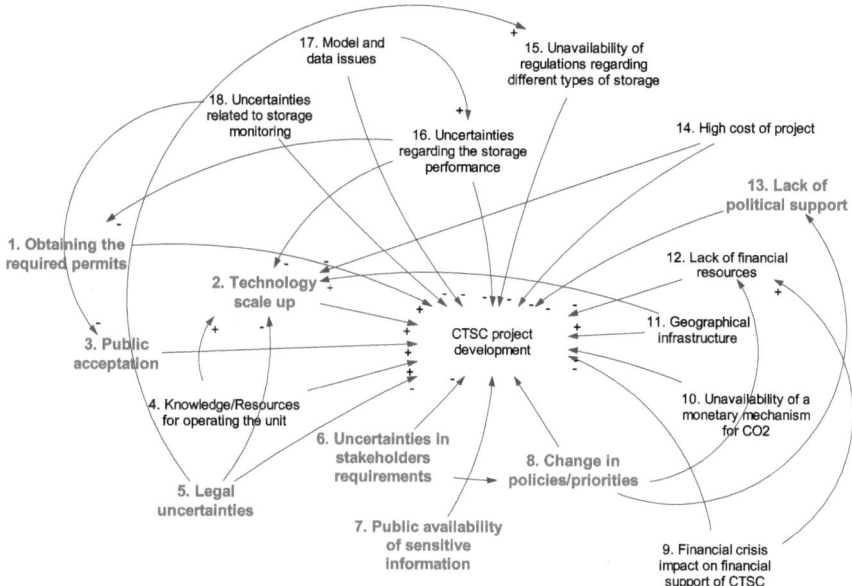

Fig. 3.16 Interconnections of major risks affecting Barendrecht project progress

Barendrecht example confirms that all potential interconnections are not identified in the risk network (Fig. 3.16). Lessons learned from the project assert that legal uncertainties/modifications, uncertainties in stakeholders' requirements and lack of political support could lead to public opposition. Hence, Fig. 3.16 should be modified as illustrated in Fig. 3.17, by adding new feedbacks.

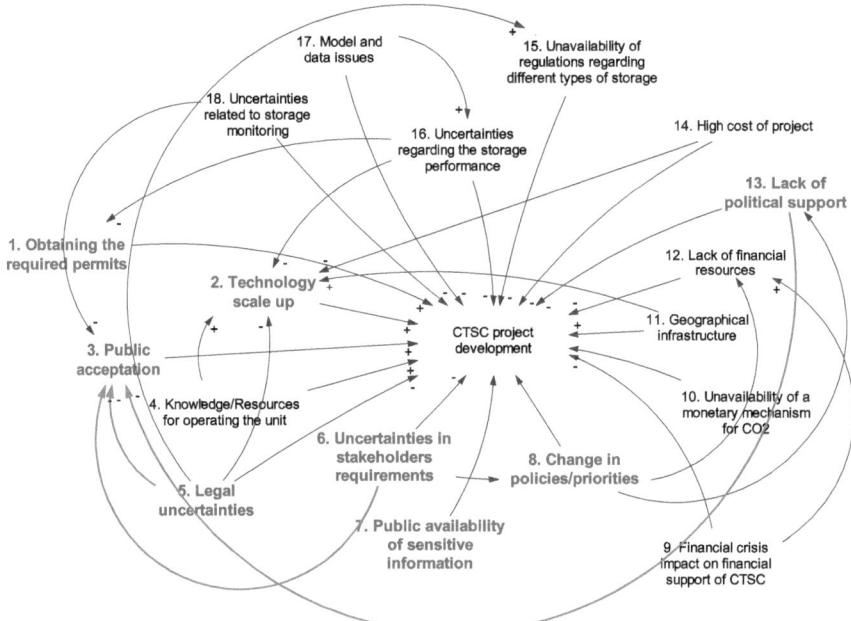

Fig. 3.17 Interconnections of major risks affecting Barendrecht project progress, modified according to lessons learned

The importance of public perception is supported by CCP (2012) that notes: *if the general public is not supportive of, or is even actively opposed to, a new technology, it can become politically and/or socially unacceptable*. CCP report also underline the role of local communities and the fact that *local communities can also create significant delays to projects, not only by influencing permitting processes, but also by physically restricting activities with demonstrations or blockades if there are significant levels of concern about a project*.

The (potential) risks involved in Lacq project context are illustrated in Fig. 3.18. The risks for which evidences/references are available are highlighted in green bold. Potential risks are represented in violet.

Figure 3.19 illustrates the (potential) risks associated to Weyburn project. Same as the previous cases, the risks for which evidences/references are available are highlighted in green bold. Potential risks are represented in violet.

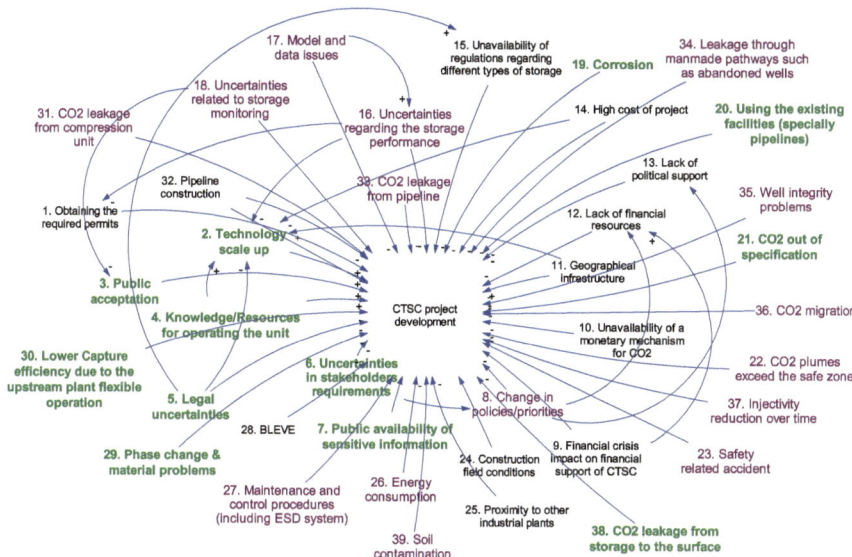

Fig. 3.18 Interconnections of major risks affecting Lacq project progress

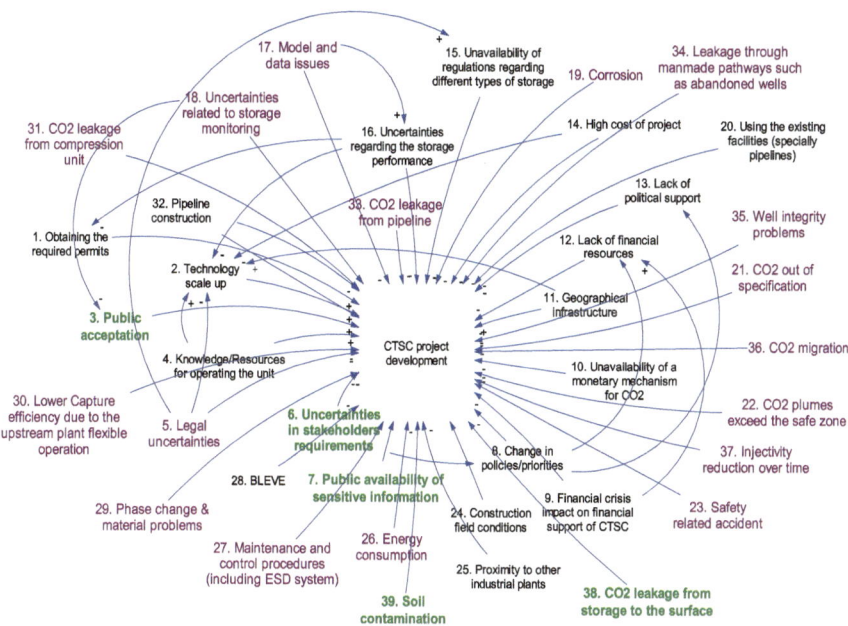

Fig. 3.19 Interconnections of major risks affecting Weyburn project progress

Weyburn case is totally different from Barendrecht and Lacq, not only due to its geopolitical context but also because Weyburn is an EOR (Enhanced Oil Recovery) project. EOR is addressed as a CO_2 reuse option rather than a long-term storage by some experts. The project is one of the Large Scale Integrated Projects which is currently in operation, even so a number of stakeholders have still some unanswered questions on the project.

The project is generally appreciated by the community. Nevertheless, there are some uncertainties supposed to be clarified by the project owners. One of the concerns, previously developed, was a leakage claim made by a farmer. Attempts were made by the project owners and independent experts to study the sources of leakage. For the moment, there is not a mutual agreement on this subject. According to available documents, local community has different opinions on the project.

Mayor of Weyburn, who has a deep familial connection to the city, is a proponent of the project. She considers Weyburn CTSC project as an opportunity for the community. She defends her idea by highlighting employment opportunities and rise in real estate business costs as positive effects of the project. The mayor believes that no safety risk is probable in long term according to the researches. Natural resources (coal) and available knowledge (on oil and gas industry) are additional points that make Weyburn an appropriate location for CO_2 storage experience (CCS101 2009a).

On the other hand, the reeve of Weyburn rural municipality is *cautiously optimistic* about the project. As well as the mayor, she has a farm family with an ancient familial background in Weyburn area. In spite of being optimistic about the project, she is *cautious* because *she doesn't feel that she knows a lot about the long-term effects. There are still some unknown factors.* The reeve makes reference to a panel organized by PTRC. She affirms that they *maybe don't have the answers that people want for those questions on long term risks.* Therefore, it is not currently obvious whether the gains from the project are short term or long term. Even if some people will come to Weyburn for working in the industry, others may leave the region because of the CO_2 storage project. The positive points are the *economic drivers* and benefits such as recovering oil (which will lead to *expand high additional employees*), media attention and tourism increase. Nevertheless, she (as both a local administration officer and a farmer) has several personal concerns. She believes that *Weyburn does rely on oil*, although agriculture is another important industry in Weyburn. Her concerns include:

– Impact of the storage on land values
– Impact of the storage on water systems
– Impact of the storage on live stock
– Impact of the storage on land production performance.

And she doubts whether Weyburn project is a long-term storage facility since oil is recovered as a result (CCS101 2009b).

These expressions attest that each stakeholder is seeking for his own individual benefits in CTSC project development. Searching for benefits (especially short-term benefits) explain why oil and gas industry is currently investing more on CTSC technologies.

Being an EOR project is a critical factor of success for Weyburn. GCCSI confirms that EOR is a significant CO_2 reuse option which has a substantial contribution to CTSC projects development. As noted before, oil production of Weyburn will increase by 130 million barrels (10% of the original oil in place) as a result of EOR operations.

3.6 A New Safety Control Structure for CTSC Projects

Reviewing the risks involved in the progress of the three case studies, direct us to the conclusion that a systematic communication among stakeholders is essential from the very first phases of the project.

Following major issues have been pointed out by CCP as the concerns of CTSC stakeholders (CCP 2007):

- Deployment cost
- Deployment scale
- Perceived risks
- Lack of accessible information
- Supporting policies
- Adequacy of regulatory frameworks.

Perceptions of several stakeholders from different geographical zones (Australia and New Zealand, North America, Europe, Japan, China, India and South Africa) are presented in this report. The stakeholders include:

- Research and Development organizations
- Industry
- Government
- NGOs
- General public

The results confirm that most of the stakeholders are worried about cost of deployment, deployment scale, impact on drinking water, accessibility of information according to the stakeholders' requirements and adequacy of regulatory frameworks in North America. However, concerns of stakeholders in Europe are much more less than the North American ones. Regulatory issues are at the top of European stakeholders' considerations. Most of the concerns have been raised by NGOs, both in North America and Europe.

The most challenging points on which there are strong difference of opinions within stakeholder groups include:

- Stakeholder perceptions on CTSC as a bridging technology
- Impact of EOR on oil market extension
- Impact of CTSC on coal market extension
- Effect on investments on other energy sources such as renewables and nuclear
- Contribution of CTSC to CO_2 emissions reduction in short term
- Inadequacy of efforts for communication
- Cost of deployment.

These points have been mostly raised in North America (CCP 2007).

Another CCP report raise below points as stakeholders' priorities (CCP 2012):

- HSE issues
- *Awareness and acceptance of CTSC*
- Technical concerns
- *Commercial and local development benefits*
- *Policy and legal issues*
- *Diversion of resources away from renewable energy*
- CTSC *positive and negative impacts on climate change*
- *Groups with variable positions on CTSC and issues of concern.*

So, the priorities are more or less similar to the previous ones.

"Conflicts about interests, values and science" is one of the factors which could create an appropriate context for emerging risks. Emerging risks may be intensified when opposition *occurs on the grounds of contested science or incompatible values*. People have subjective views about the science according to their own values. Hence, in case of conflicts, interests and values of involved stakeholders should be clarified. Examples are available for both successful and failed attempts to *block a technology or industrial facility*. The positive one is the conflicts on potential risks of LNG terminals, which are managed successfully in the Netherlands *through creative use of public participation and local discussion*. In the contrary, the US nuclear waste management is a *failed* example.

"Social dynamics" is another critical factor. *Societies are continually evolving. As complex systems, they may adapt to new or changing technologies ... However, they sometimes fail to adapt.* Social dynamics are not *directly controllable* but may be *influenced* in order to mitigate emerging risks (IRGC 2010).

Internal and external communication can also affect emerging risks intensification; internal communication between the actors involved in risk management, and external communication of these actors with the public. IRGC report underlines varied concerns of people and scientists/regulators concerning CO_2 Capture and Storage. Some people are worried about safety risks and ground water contamination while others are more concerned about the cost, *the effect on their electric rates* and *property values* (IRGC 2010). Communication allows improving risk management process by integrating all stakeholder concerns.

Analysis of the case studies result in the following general safety control structure (Fig. 3.20). The Figure confirms the importance of communication among stakeholders.

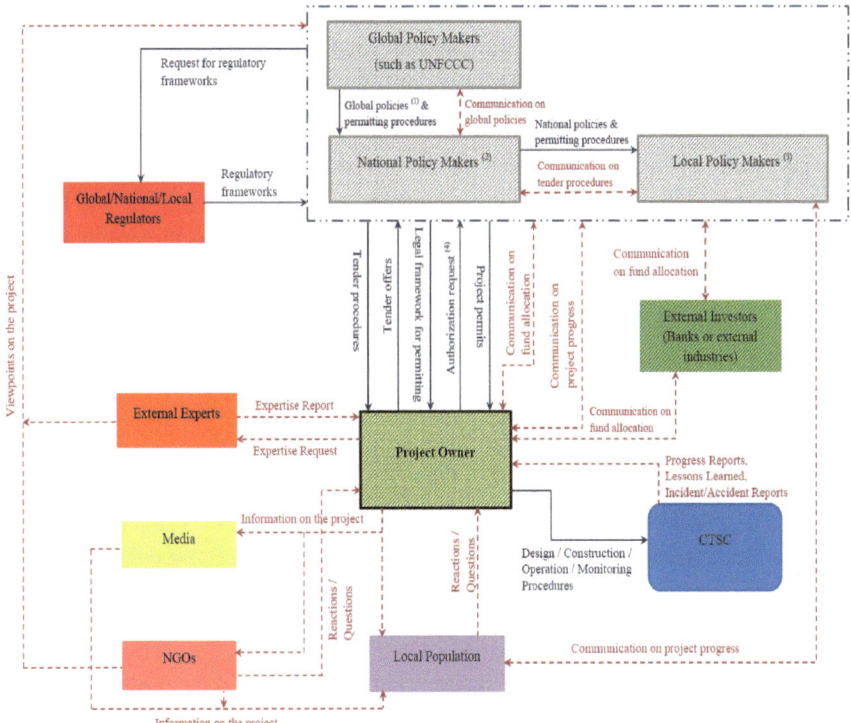

Fig. 3.20 Proposed safety control structure for CTSC projects. (1) Global policies according to regulatory frameworks. (2) Including Policy Makers in the scale of zones (EU, US, etc.) and countries. (3) Including Policy Makers in the scale of regions and communities. (4) Including EIA, Hazard Analysis and HSE concerns

LEGEND

☐	Stakeholders (Controllers)	→	Documents and information, not necessarily exchanged dynamically
☐	Physical Plant	╌╌→	Dynamically interchanged documents / actions
■	Potential Investors		

In Fig. 3.20, solid lines represent documents and information exchanged between the stakeholders, not necessarily in a dynamic manner. Dash lines show the flow of dynamic interchange, i.e. what should be maintained throughout the project life.

Global, National and Local Governments are regrouped in a box, since the relationship of other stakeholders with the governments is varied in different regions.

Regulators are asked by Policy Makers for regulatory frameworks. Global policies and permitting procedures are defined (by Global Policy Makers) for CTSC according to regulatory frameworks and climate change policies. National policies and permitting procedures are transposed to national contexts by National Policy Makers, who shall be continuously in communication with Global and Local Policy Makers.

Tender procedures are sent to the Project Owner by the government. The Project Owner returns the tender offers and if the offer is accepted, project permits will be provided in reply to the authorization request of the Project Owner.

The hatched squares (including Governments, Project Owner and External Investors) represent potential investors of the project who should intercommunicate on the funds allocated for the project. There are still several uncertainties about the actors who have to pay for developing CTSC technologies.

External Experts are always engaged to provide expertise usually on technical aspects of the project.

Information on the project has to be shared dynamically with all stakeholders including Local Population, NGOs and Media.

Communication is also essential between governments, NGOs and Local Population, since local communities need to be assured of political support of their policy makers in order to accept CTSC as a novel beneficial technology.

Delays, especially in communication, have to be minimized.

Figure 3.20 underlines the significance of information feedback loops within the safety control structure of CTSC projects. As discussed previously in Chap. 2, information feedbacks allow the actors to improve their mental models, decisions, strategies and decision rules.

Dulac asserts this opinion by remarking that *improving mental models will consequently improve the quality of safety-related decision-making … and the performance of organizations and systems* (Dulac 2007; Leveson 2009). As previously mentioned, risk acceptance and risk communication are integrated in risk management process (Condor et al. 2011). Risk communication involves providing information for stakeholders to improve their understanding of the risks related to a phenomenon or a technology. Mental models are the schemas of human beings which help them make decisions. Investigating mental models of both experts and lay people provide essential information for communication (Skarlatidou et al. 2012).

References

CBC news (2011) CO$_2$ leaks worry Sask. Farmers, last updated: Tuesday, January 11, 2011, 8:49 PM. http://www.cbc.ca/news/business/story/2011/01/11/sk-carbon-complaint-1101.html. Accessed 10 Sept 2012

CCJ (2010) Shell Barendrecht project cancelled, Carbon Capture Journal, November 5, 2010. http://www.carboncapturejournal.com/displaynews.php?NewsID=676. Accessed 22 June 2012

CCP (2007) Public perception of carbon dioxide capture and storage: prioritized assessment of issues and concerns. Summary for Policy-Makers, CO_2 Capture Project®, March 2007. http://ccs101.ca/assets/Documents/iea_public_perception_of_ccs.pdf. Accessed 20 July 2012

CCP (2010) Update on selected regulatory issues for CO_2 Capture and Geological Storage, CO_2 Capture Project®. Final report, Nov 2010

CCP (2012) CCS stakeholder issues, Review and Analysis, CO_2 Capture Project®. Final report, Feb 2012

CCS101 (2009a) Interview with Debra Button, Mayor of Weyburn, October 2009, The carbon capture and storage information source. http://ccs101.ca/ccs_communities. Accessed 26 Sept 2012

CCS101 (2009b) Interview with Carmen Sterling, Reeve of the Rural Municipality of Weyburn, October 2009. The carbon capture and storage information source. http://ccs101.ca/ccs_communities. Accessed 26 Sept 2012

CLIS (2008) CLIS (Commission Locale d'Information et de Suivi), Évaluation du projet de captage et stockage de CO_2 à Rousse 1, Dossier présenté par Total, 29/09/2008. http://www.pyrenees-atlantiques.pref.gouv.fr/sections/actions_de_l_etat/environnement_et_dev/actualites/. Accessed 12 Nov 2010

CLIS (2011) CLIS (Commission Locale d'Information et de Suivi), Demande de prolongation de la phase d'injections, 12/09/2011. http://www.pyrenees-atlantiques.pref.gouv.fr/sections/actions_de_l_etat/environnement_et_dev/actualites/. Accessed 15 Dec 2011

Condor J, Unatrakarn D, Wilson M, Asghari K (2011) A comparative analysis of risk assessment methodologies for the geologic storage of carbon dioxide. Energy Procedia 4(2011): 4036–4043. https://doi.org/10.1016/j.egypro.2011.02.345

C&S Conseils (2008) Bilan de la concertation, Projet pilote de captage stockage de CO_2 dans le bassin de Lacq, Janvier 2008. http://www.total.com/MEDIAS/MEDIAS_INFOS/2186/FR/. Accessed 13 Apr 2012

Dulac N (2007) A framework of dynamic safety and risk management modeling in complex engineering systems. PhD thesis submitted to the department of aeronautical and astronautical engineering at MIT, Feb 2007

Ecojustice (2011) Ecojustice reply to PTRC response to soil gas studies, January 28, 2011. http://www.ipac-co2.com/uploads/File/PDFs/ecojustice%20response.pdf. Accessed 20 July 2012

Feenstra CFJ, Mikunda T, Brunsting S (2010) What happened in Barendrecht? Case study on the planned onshore carbon dioxide storage in Barendrecht, the Netherlands, ECN/CAESAR, 3 Nov 2010

GCCSI (2009) Strategic analysis of the global status of carbon capture and storage. Report 5: synthesis report, Global CCS Institute, Canberra, Australia

GCCSI (2011) The global status of CCS: 2011. Global CCS Institute 2011, Canberra, Australia. ISBN 978-0-9871863-0-0

GCCSI (2012) Global status of large scale integrated CCS projects, June 2012 update

GCCSI (2016) The global status of CCS: 2016, Global CCS Institute 2016, Summary Report, Australia, ISBN 978-0-9944115-6-3

GCCSI website (2017) Global CCS Institute, Projects status. https://www.globalccsinstitute.com/projects/large-scale-ccs-projects. Accessed 27 Oct 2017

ICPE website 2. http://www.installationsclassees.developpement-durable.gouv.fr/Qui-controle.html. Accessed 5 Sept 2012

ICPE website 3. http://www.installationsclassees.developpement-durable.gouv.fr/Quelle-procedure-suivra-la-demande.html. Accessed 5 Sept 2012

IRGC (2010) The emergence of risks: contributing factors. Report of International Risk Governance Council, Geneva, 2010. ISBN 978-2-9700672-7-6

Kapetaki Z, Scowcroft J (2017) Overview of Carbon Capture and Storage (CCS) demonstration project business models: risks and enablers on the two sides of the Atlantic. Energy Procedia 114(2017):6623–6630. https://doi.org/10.1016/j.egypro.2017.03.1816

Kerlero de Rosbo G (2009) Integrated Risk Analysis for large-scale CCS projects implementation, Thèse professionnelle du Master Spécialisé en Ingénierie et gestion de l'Environnement (ISIGE)

Lacq Project (2012) Documents concerning Lacq Project, available on the regional prefecture website. http://www.pyrenees-atlantiques.pref.gouv.fr/Politiques-publiques/Environnement-risques-naturels-et-technologiques/Pilote-d-injection-de-CO2/Projet-de-captage-CO2-de-Total/. Accessed from Nov 2010 to Oct 2012

Lafleur P (2011) Geochemical soil gas survey, A Site Investigation of SW30-5-13-W2M Weyburn Field, SASKATCHEWAN, Monitoring Project Number 2, Petro-Find Geochem Ltd, Submitted to: Cameron and Jane Kerr, March 16, 2011. http://www.ipac-co2.com/uploads/File/PDFs/PetroFind_Report_Kerr_Feb_2011_Final.pdf. Accessed 20 July 2012

Leveson N (1995) Safeware, system safety and computers. Addison-Wesley Publishing Company. ISBN 0201119722

Leveson N (2004) Model-based analysis of socio-technical risk, Massachusetts Institute of Technology, Engineering Systems Division. Working Paper Series, ESD-WP-2004-08, Dec 2004

Leveson N (2009) Engineering a safer world, system safety for the 21st century, Massachusetts Institute of Technology, July 2009 (a draft book). http://sunnyday.mit.edu/safer-world.pdf. Accessed 1 Dec 2011

Longannet (2011) FEED documents of Longannet project, published online http://www.decc.gov.uk/en/content/cms/emissions/ccs/demo_prog/feed/scottish_power/scottish_power.aspx. Accessed 29 Nov 2011

Mairie Jurançon (2012) http://www.annuaire-mairie.fr/statistique-jurancon.html. Accessed 10 Sept 2012

MIT (2016) Massachusetts Institute of Technology, Cancelled or Inactive Projects. http://sequestration.mit.edu/tools/projects/index_cancelled.html. Accessed 4 July 2017

Netherlands Government (2010) http://www.rijksoverheid.nl/nieuws/2010/11/04/co2-opslagproject-barendrecht-van-de-baan.html. Accessed 9 Aug 2012

PMBOK (2008) A guide to the Project Management Body of Knowledge (PMBOK® Guide), 4th edn, Project Management Institute, USA. ISBN 978-1-933890-5-7

Poumadère M, Bertoldo R, Samadi J (2011) Public perceptions and governance of controversial technologies to tackle climate change: nuclear power, carbon capture and storage, wind and geoengineering. John Wiley & Sons, WIREs Climate Change 2011. https://doi.org/10.1002/wcc.134

PTRC (2004) IEA GHG Weyburn CO_2 Monitoring & Storage Project: Summary Report 2000-2004. In: From the proceedings of the 7th international conference on greenhouse gas control technologies, 5–9 September 2004, Vancouver, Canada

Rochon E, Kuper J, Bjureby E, Johnston P, Oakley R, Santillo D, Schulz N, Von Goerne G (2008) False hope: why carbon capture and storage won't save the climate, Greenpeace, Published in May 2008 by Greenpeace International, Amsterdam, The Netherlands

Skarlatidou A, Cheng T, Haklay M (2012) What do lay people want to know about the disposal of nuclear waste? A mental model approach to the design and development of an online risk communication. Risk Anal 32(9). https://doi.org/10.1111/j.1539-6924.2011.01773.x

Soupa O, Lajoie B, Long A, Alvarez H (2012) Leading the energy transition: bringing carbon capture and storage to market. Schlumberger and SBC Energy Institute, June 2012

Thomas N, Last G, Robertson A, Goldsmith J (2012) Carbon capture and storage: lessons from the competition for the first UK demonstration, National Audit Office, 16 March 2012, London, UK

Tombari J (2011) Ten uncertainties around CO_2 storage costs, GCCSI blog, 05 April 2011. http://www.globalccsinstitute.com/community/blogs/authors/tombari1/2011/04/05/ten-uncertainties-around-co2-storage-costs. Accessed 1 Aug 2012

Verdon JP (2012) Microseismic monitoring and geomechanical modelling of CO_2 storage in subsurface reservoirs, Springer theses, Chapter 2, The Weyburn CO_2 Injection Project. ISBN 978-3-642-25387-4. www.springer.com/.../9783642253874-c2.pdf? Accessed 19 July 2012

Whittaker S (2011) PTRC response to a soil case study performed by petro-find geochem LTD., submitted to the Ministry of Energy and Resources, Petroleum and Natural Gas Division, Regina, January 19, 2011. http://www.ipac-co2.com/uploads/File/PDFs/whittaker%20letter%20to%20danscok%2019jan2011%20with%20ptrc%20response.pdf. Accessed 20 July 2012

Conclusion

Overview of the Proposed Methodology

Capture, Transport and Storage of CO_2 (CTSC) is considered as an essential technology for climate change mitigation. However, risks and uncertainties related to long term reliability of the technology have resulted in a kind of uncertain future for CTSC projects development.

CTSC is claimed to play *a new moderating role in opposition to coal* (Stephens 2012). Such moderating role is extremely important in the current coal-dependent energy policy. On the other hand, CTSC has been sometimes expressed as a technology that leads to *fossil-fuel lock-in* (Unruh and Carrillo-Hermosilla 2006; Vergragt et al. 2011). It is argued that CTSC will not help getting rid of fossil fuels. On the contrary, it could amplify the dependence of energy market on fossil fuels. Stephens believes that CTSC deals with a two-fold lock-in: technical and political. She argues that *for those governments and private companies that have already invested millions or billions of dollars to advance CCS, ending their support for this technology may be difficult even if perceptions of the relative challenges and potential of CCS continues to change over time* (Stephens 2012).

Koornneef et al. have identified several knowledge gaps in the field of CTSC environmental and risk assessment, which *may have the potential to postpone the implementation of CCS.* They believe that uncertainties regarding risk assessment could be a *bottleneck for wide scale implementation of CCS if not properly addressed.* In terms of technical risk assessment, Capture and Transport are supposed to be sufficiently understood, although further studies are required to identify potential failure scenarios and their consequences. CO_2 storage is known as a *non-engineered* part of the chain for which quantitative risk assessment is currently impossible (Koornneef et al. 2012). EU commission has confirmed that uncertainty is a major barrier to invest on low carbon energy systems (EU commission 2011).

A systemic risk management framework for CTSC projects has been proposed in this work. The approach is founded on the concepts of systems thinking, STAMP, STPA and system dynamics. The objective is to provide a means of decision

© The Author(s) 2018
J. Samadi and E. Garbolino, *Future of CO₂ Capture, Transport and Storage Projects*,
SpringerBriefs in Environmental Science, https://doi.org/10.1007/978-3-319-74850-4

making for CTSC projects development in the actual context where the future of the technology is uncertain. Risk management is considered as a means of control that should be able to propose a control structure for the whole system. Stakeholders are viewed as controllers of the system. Four conditions are required for the controller (Leveson 2009):

– Having a goal
– Being able to affect the system
– Being or contain a model of the system
– Being able to observe the system

A number of projects have been cancelled or delayed for various reasons. Refer to available information forty-three projects have been cancelled or put on hold all over the world. Financial reasons are frequently noted as the reason of project failure. Nevertheless, public opposition and legal issues are other causes of project cancellation (MIT 2016).

In order to analyze the risks preventing project progress, the ones related to the phases prior to engineering have been selected and modeled by the proposed methodology. The aim was to study the feedback networks affecting the risks amplification. The analysis has been started from stock/flow models of each risk. Models have been subsequently grouped together in order to study interconnections of risks and feedback loops result in project failure or success.

Safety control structures of three case studies have been reviewed to find a generic structure that could work for CTSC projects. Inadequate control actions to maintain safety constraints have been discussed. The idea comes from STAMP and STPA approaches, developed at MIT. The purpose was to underline the significance of endogenous point of view in analyzing the risks of CTSC projects. It has been argued that feedbacks and feedback loops have to be understood and studied in the networks of risks and stakeholders. Emphasis is placed on the importance of providing endogenous explanations for CTSC actual development context.

CTSC risk management is context specific and depends on several factors such as national and local circumstances. In spite of that, seeking for individual benefits is indeed a major concern of all stakeholders. Oil and gas industry is currently more involved in the field by investing on CTSC EOR projects. Oil recovery increase is the main obvious advantage of EOR systems.

Lessons learned from the modeling process of this work show that dynamic information sharing and communication are essential to support the contribution of CTSC technologies in climate change mitigation.

This work provides a decision-making support for the progress of CTSC projects. Systemic modeling of CTSC project risks can help the stakeholders to share and improve their mental models and accordingly, their strategies and decisions.

Advantages of the Methodology

In order to give a summary of the proposed methodology advantages, we have to go back to available CTSC risk management approaches. As discussed in Chap. 1, several works have been already performed on risk management of CTSC. Most of these works are focused on one part of the chain, i.e. Capture, Transport or Storage; and especially on technical aspects of risk. However, some integrated approaches are available for CTSC risk management. INERIS, National Institute of Industrial Environment and Risks in France, proposes a global risk analysis approach for CTSC chain. They propose to integrate the notion of time to the classic concepts of probability and severity for CTSC risk analysis. Three time scales are suggested: operation (max. 50 years), monitoring (max. 150–200 years) and long term (up to 1000 years). Different aspects of risks are not included in the approach of INERIS. Their study is focused on technical risk scenarios related to storage (Farret et al. 2009). Therefore, in subsequent paragraphs we will review the main characteristics of two available integrated approaches for the purpose of better understanding the values of our proposed systemic methodology.

GCCSI has presented a qualitative risk assessment methodology which has been developed based on AS/NZS 4360: 2004 (Australian and New Zealand standard for risk management). Seventeen extreme risks have been identified by an expert panel, and classified in four main categories: Public, Business Case, Governmental/ Regulatory/Policy and Technical. Consequences and likelihood of each risk have been then specified by the expert panel (GCCSI 2009).

GCCSI asserts that many of these risks are complex, inter-related and dynamic (GCCSI 2009). Nevertheless, the complexity, interrelations and dynamic characteristic of risks have not been studied by GCCSI. Therefore, the advantage of our proposed methodology compared with GCCSI approach is that our systemic methodology provides a modeling framework for analyzing the complex interrelation network of risks associated to CTSC projects. In addition to GCCSI (2009), a number of recent references have been used to determine the risk categories of the present work (Table 3.1). Consequently, our risk categories are more comprehensive than the ones presented by GCCSI.

Another integrated risk assessment approach has been proposed by Kerlero de Rosbo (2009) for Belchatow project in Poland. Risks have been sorted out in five main groups: Technical, Financial, Organization & Management, Social & Political, and Regulatory. A semi-quantitative approach has been applied by the author. The methodology steps are indeed same as a classic risk management process, including analysis, evaluation and treatment of risks.

Risks as well as their likelihood and severity have been identified in expert panels. Although several aspects of risk have been included in Kerlero's methodology, interconnections of risks are not analyzed in his approach.

In addition to risk interrelations, another point which seems to be necessary to be integrated in CTSC risk management processes is the importance of stakeholders role in the project success or failure. The significance of safety control structure (as defined in Chap. 3) has not been taken into account in the integrated methods of (GCCSI 2009; Kerlero de Rosbo 2009). Responsibilities of different stakeholders of CTSC project is what we have highlighted in our systemic approach. Each stakeholder is considered as a controller who has to maintain specific safety constraints in order to fulfill the objective of safety structure, i.e. preventing delay or failure of CTSC project. In the current work, defects of safety control structure have been noted as major potential cause of a CTSC project failure.

To sum up, three advantages can be listed for the systemic methodology which is proposed in this book:

– Presenting more comprehensive list and categories of risks related to CTSC chain
– Taking into account the complex network of risk interconnections by proposing a systemic modeling framework
– Underlining the significance of stakeholders' role in the project success or failure, by proposing a modeling approach for safety control structure of projects and analyzing required and (potential) inadequate control actions of stakeholders in relation to each risk

By the way, this approach has also some limitations in spite of its advantages and added values. Limitations are classified in three groups presented hereafter.

Limitations of the Methodology

Lack of Information on CTSC

CTSC integrated chain is an emerging technology for which there is not a great amount of publicly available information. Details of case studies are usually unavailable due to confidentiality issues. Nevertheless, the methodology has been applied for three case studies on the basis of accessible data in the literature, project reports and discussions with experts. The analysis could be improved based upon lessons learned from further development of projects.

Qualitative Versus Quantitative Approach

A qualitative approach was proposed in this work for risk management of CTSC. It may be debated that quantitative methods are more practical or more comprehensible. In this section, the notion of quantification is reviewed from three points of view: risk quantification, quantification in STAMP approach, and system dynamics quantitative modeling.

As mentioned earlier in Chap. 1, *operators and public organizations have initially tried to quantify damages and consequences of potential accidents, before to understand why and how they could occur* (Tixier et al. 2002). From another standpoint, quantitative approaches are not necessarily the most adapted ones for modern complex sociotechnical systems (Dulac 2007). Altenbach mentions ten reasons for which risks should not be quantified. Controversiality, potential use of numbers out of context, simplification of numbers for challenge and criticism, being *time consuming and costly, uncertainties,* requirement of *more training, data requirement, being threatening and compelling, usefulness of qualitative results and difficulty to communicate the concept of probability* are noted as the reasons not to quantify risks (Altenbach 1995).

Our proposed methodology is based on STAMP approach, which has been mostly used as a qualitative tool to analyze accidents or risks. Dulac affirms that *quantitative values generated in the simulations are sometimes of secondary importance in comparison to the qualitative learning opportunities presented by the model and the modeling process* (Dulac 2007). The significance of modeling process is also attested by Durand (2010).

From system dynamics point of view, qualitative or *"soft"* applications of stock-flow and/or causal diagrams are recognized as useful as simulation applications. Qualitative use allows developing feedback networks and understanding the system behavior (Winch 2000).

Hence, being qualitative is not a limitation of the proposed methodology. As Coyle suggests, we should wonder *how much value does quantified modeling* in system dynamics *add to qualitative analysis* (Coyle 2000).

Recent studies compare qualitative and quantitative risk assessment methods for CTSC, stating that the use of quantitative risk assessment methods at this point is challenging because of a lack of specific data. However, the development of frameworks and qualitative methods might be the most trustworthy for current projects (Wennersten et al. 2015).

Subjectivity of Modeling and Risk Assessment

Modeling, which is a simplification of reality, is made by an individual or a group of individuals. As a result, modeling is always a subjective process, depending on the reasoning of modeler(s). The models developed in this work are not an exception. They have been created based on the mental models of the modeler, which are inevitably restricted. According to Durand, modeling is an art and not an established technique (Durand 2010). Models of the current work are made by only one modeler and have not been verified by an expert panel. Group modeling provides different points of view to improve the models.

In addition, risk assessment is a subjective process since expert judgment is an indispensable characteristic of risk assessment process.

Final Word

Stakeholders have still different positions on CTSC technology and its commitment on Climate Change.

The leaders of ten major oil and gas companies have recently announced that they aim to help make CTSC a *commercial reality*. This is part of their aim to *make the world's energy systems fit for the future* (OGCI 2017)

While Greenpeace, as a NGO against CTSC, believes that CTSC has not advanced much since their first study in 2008 (Rochon et al. 2008). Many projects have been cancelled in recent years because of high costs and technical issues. Instead of investing on such an *expensive and risky distraction*, it is better to invest on renewables (Greenpece 2016).

Experts argue that CTSC is still an unknown technology for many stakeholders. In order to develop CTSC projects, transparent communication process is necessary about different aspects of risks and the mitigation options.

Risk communication must start in the very first phases of the project and involve all stakeholders.

References

Altenbach TJ (1995) A comparison of risk assessment techniques from qualitative to quantitative. In: Submitted to the ASME pressure and piping conference, Honolulu, Hawaii, 23–27 July 1995

Coyle G (2000) Qualitative and quantitative modelling in system dynamics: some research questions. Syst Dyn Rev 3(16):225–244

Dulac N (2007) A framework of dynamic safety and risk management modeling in complex engineering systems. PhD thesis submitted to the department of aeronautical and astronautical engineering at MIT, Feb 2007

Durand D (2010) La systémique, Presses Universitaires de France (PUF), 11th edn, Jan 2010

EU Commission (2011) Communication from the commission to the European parliament, the council, the European economic and social committee and the committee of the regions, Energy Roadmap 2050, European Commission. http://ec.europa.eu/energy/energy2020/roadmap/doc/com_2011_8852_en.pdf. Accessed 30 Jan 2012

Farret R, Gombert P, Lahaie F, Roux P (2009) Vers une méthode d'analyse des risques globale de la filière CSC, intégrant plusieurs échelles de temps, Rapport Scientifique, INERIS, 2008-2009, pp 100–103

GCCSI (2009) Strategic analysis of the global status of carbon capture and storage. Report 5: synthesis report, Global CCS Institute, Canberra, Australia

Greenpeace (2016) Carbon capture and storage a costly, risky distraction, 1 July 2016. http://www.greenpeace.org/international/en/campaigns/climate-change/Solutions/Reject-false-solutions/Reject-carbon-capture–storage/. Accessed 7 Nov 2017

Kerlero de Rosbo G (2009) Integrated risk analysis for large-scale CCS projects implementation. Thèse professionnelle du Master Spécialisé en Ingénierie et gestion de l'Environnement (ISIGE)

Koornneef J, Ramirez A, Turkenburg W, Faaij A (2012) The environmental impact and risk assessment of CO_2 capture, transport and storage—an evaluation of the knowledge base. Prog Energy Combust Sci 38:62–86. https://doi.org/10.1016/j.pecs.2011.05.002

Leveson N (2009) Engineering a safer world, system safety for the 21st century, Massachusetts Institute of Technology, July 2009 (a draft book). http://sunnyday.mit.edu/safer-world.pdf. Accessed 1 Dec 2011

MIT (2016) Massachusetts Institute of Technology, Cancelled or Inactive Projects. http://sequestration.mit.edu/tools/projects/index_cancelled.html. Accessed 4 July 2017

OGCI (2017) Catalyst for change: collaborating to realize the energy transition. A report from the oil and gas climate initiative, Oct 2017

Rochon E, Kuper J, Bjureby E, Johnston P, Oakley R, Santillo D, Schulz N, Von Goerne G (2008) False hope: why carbon capture and storage won't save the climate, Greenpeace, Published in May 2008 by Greenpeace International, Amsterdam, The Netherlands

Stephens JC (2012) An uncertain future for carbon capture and storage (CCS). Federation of American Scientists, 28 June 2012. http://www.fas.org/blog/pir/2012/06/28/an-uncertain-future-capture-and-storage-for-carbon-css/. Accessed 22 Aug 2012

Tixier J, Dusserre G, Salvi O, Gaston D (2002) Review of 62 risk analysis methodologies of industrial plants. J Loss Prev Process Ind (15):291–303

Unruh GC, Carrillo-Hermosilla J (2006) Globalizing carbon lock-in. Energy Policy 34: 1185–1197. https://doi.org/10.1016/j.enpol.2004.10.013

Vergragt PhJ, Markusson N, Karlsson H (2011) Carbon capture and storage, bio-energy with carbon capture and storage, and the escape from the fossil-fuel lock-in. Global Environ Change 21:282–292. https://doi.org/10.1016/j.gloenvcha.2011.01.020

Wennersten R, Sun Q, Li H (2015) The future potential for carbon capture and storage in climate change mitigation—an overview from perspectives of technology, economy and risk. J Cleaner Prod 103:724–736

Winch GW (2000) System dynamics: from theory to practice. In: 1st international conference on systems thinking in management